Toward High School Biology

Understanding Growth in Living Things

Teacher Edition

NSTApress®
National Science Teachers Association

Claire Reinburg, Director
Rachel Ledbetter, Managing Editor
Deborah Siegel, Associate Editor
Amanda Van Beuren, Associate Editor
Donna Yudkin, Book Acquisitions Manager

Art and Design
Will Thomas Jr., Director

Printing and Production
Catherine Lorrain, Director

National Science Teachers Association
David L. Evans, Executive Director
David Beacom, Publisher

1840 Wilson Blvd., Arlington, VA 22201
www.nsta.org/store
For customer service inquiries, please call 800-277-5300.

20 19 18 17 4 3 2 1

NSTA is committed to publishing material that promotes the best in inquiry-based science education. However, conditions of actual use may vary, and the safety procedures and practices described in this book are intended to serve only as a guide. Additional precautionary measures may be required. NSTA and the authors do not warrant or represent that the procedures and practices in this book meet any safety code or standard of federal, state, or local regulations. NSTA and the authors disclaim any liability for personal injury or damage to property arising out of or relating to the use of this book, including any of the recommendations, instructions, or materials contained therein.

Library of Congress Cataloging-in-Publication Data
Names: National Science Teachers Association.
Title: Toward high school biology : understanding growth in living things.
Other titles: Understanding growth in living things
Description: Teacher edition. | Arlington, VA : National Science Teachers Association, 2017.
Identifiers: LCCN 2017021524 (print) | LCCN 2017025654 (ebook) | ISBN 9781681405612 (e-book) | ISBN 9781681405605 (print)
Subjects: LCSH: Growth--Juvenile literature. | Growth--Study and teaching (Secondary)
Classification: LCC QH511 (ebook) | LCC QH511 .T69 2017 (print) | DDC 571.8--dc23
LC record available at https://lccn.loc.gov/2017021524

SUSTAINABLE FORESTRY INITIATIVE Certified Sourcing
www.sfiprogram.org
SFI-00756

Contents

Note: Color images of models are available on the book's Extras page at *www.nsta.org/towardhsbio*.

AAAS Project 2061 Team
Jo Ellen Roseman, Principal Investigator
Cari Herrmann-Abell, Senior Research Associate
Jean Flanagan, Research Associate
Ana Cordova, Research Assistant
Caitlin Klein, Research Assistant
Bernard Koch, Research Assistant
Mary Koppal, Communications Director
Abigail Burrows, Senior Project Coordinator

External Review Team
Marlene Hilkowitz, Temple University, Philadelphia, PA
Michele Lee, Temple University, Philadelphia, PA
Barb Neureither, Michigan State University, East Lansing, MI
Edward Smith, Michigan State University, East Lansing, MI

Consultant
Kathy Vandiver, Massachusetts Institute of Technology, Boston, MA

Acknowledgments
The *Toward High School Biology* unit has benefited immensely from the many contributions of the staff at BSCS, who worked in partnership with AAAS during the first three years of the project (2010 through 2013). In particular, we are grateful to Janet Carlson, Rhiannon Baxter, Brooke Bourdélat-Parks, Elaine Howes, Rebecca Kruse (currently at the National Science Foundation), Stacey Luce, Chris Moraine, Kathleen Roth, and Kerry Skaradzinski for their expertise and tireless efforts. Recent contributions (2014 to 2016) of Rebecca Kruse were funded by the National Science Foundation Independent Research/Development Program. Any opinions, findings, conclusions, or recommendations expressed in this publication are those of the authors and do not necessarily reflect the views of the National Science Foundation.

We would also like to thank the many excellent teachers and their students in Colorado, the District of Columbia, Maryland, and Massachusetts who participated in pilot- and field-testing the unit. Their insights and feedback have been invaluable.

Development of *Toward High School Biology* was funded by the U.S. Department of Education Institute of Education Sciences, Grant #IES-R305A100714. Any opinions, findings, conclusions, or recommendations expressed in this publication are those of the authors and do not necessarily reflect the views of the funding agency.

INTRODUCTION

About the Teacher Edition

The Teacher Edition is designed to provide easy access to essential background information and support needed for using the *Toward High School Biology* (*THSB*) unit effectively in the classroom. Additional teacher support materials (e.g., brief video demonstrations, color images from the unit, and tutorials) are provided on the book's Extras page at *www.nsta.org/towardhsbio*. The Teacher Edition and the online materials aim to provide both a "big picture" sense of the unit and its goals as well as the specific information and guidance needed to teach each lesson and carry out each activity.

Features of the Teacher Edition

At the unit level, the **Unit Overview** describes the overarching goal of the unit as it relates to the central questions students should be able to answer and the core disciplinary ideas they should understand after completing the unit. **Alignment With National Science Education Framework and Standards** describes the disciplinary core ideas, science practices, and crosscutting concepts from the National Research Council's *A Framework for K–12 Science Education* (the *Framework*; NRC 2012) around which the unit is developed and the performance expectations from *Next Generation Science Standards* (*NGSS*; NGSS Lead States 2013) that the unit contributes to. **Structure of the Unit** describes and **Content Storyline** represents where in the unit specific science ideas are developed and how the ideas build on each other from chapter to chapter. **Unique Aspects of the Unit** describes the rationale for the unit's approach to the science content, including its use of science terminology in the Student Edition. **Student Edition Conventions** presents the design principles guiding the development of lessons and describes the purpose and key features of each section of a lesson.

At the chapter level, the **Chapter Overview** describes the concepts developed in that chapter and provides a short synopsis of what students will do and think about in each lesson. **Background Knowledge for Teachers** provides information about key ideas to be developed in the chapter, as well as what phenomena, data, and models may be used to develop or illustrate those ideas. When appropriate, intended observations and rationale for selection are discussed. **Prerequisite Knowledge for Students** describes assumptions about where students are likely to be in their understanding of the science ideas that are central to each chapter and identifies knowledge from other disciplines (e.g., English and Language Arts) that students are expected to have. **Common Student Ideas and Misconceptions** outlines a small number of common confusions and misunderstandings from the research literature and from our own experiences in previous work that may be encountered in teaching the unit.

At the lesson level, each lesson begins with a **Lesson Guide** that provides an overview of the lesson, including the key question, targeted sciences ideas and practices, materials and advance preparation needed, and a summary chart describing key phenomena, data and models used, intended observations, and the pedagogical purpose of each. The remaining pages of the lesson consist of facing **Student Edition Page and Answer Key** and **Teacher Facilitation Notes** pages. Student Edition pages are identified as "**SE p. 1**," "**SE p. 2**," and so on so that the page numbers are consistent with those that students see in their workbooks. Each Student Edition Page and Answer Key page includes the Student Edition page plus ideal student responses (in gray text) written to reflect what students are likely to understand at that point in the lesson. Each page of the Teacher Facilitation Notes has a header that corresponds to a header on the facing student page and includes a time estimate in parentheses. The time each activity takes may vary according to each classroom's needs. Each lesson concludes with the heading *Closure and Link*, which is only in the Teacher Edition. Closure and Link is a description of the discussion that should arise at the end of each lesson/chapter. Ideally, students would come to the listed ideas and conclusions through a facilitated discussion. Teacher Facilitation Notes also include *Teacher Talk and Actions*, strategies for facilitating each page of the Student Edition, including partner/small group work, prompts for whole-class discussion, issues/ideas to highlight, and additional Science Notes for the teacher to be aware of. The Teacher Facilitation Notes are not a script; instead the Notes provide the essence of the developers' intent and suggest reasonable ways to adjust the curriculum to address the needs of particular students or classroom situations while still maintaining consistency with that intent.

Unit Overview

In developing the *Toward High School Biology: Understanding Growth in Living Things* unit, the American Association for the Advancement of Science (AAAS) and Biological Sciences Curriculum Study (BSCS) team was guided by learning theory and research evidence showing that students' science understanding develops from (a) having a wide range of experiences with the natural world that are explainable by a coherent set of ideas and (b) having an opportunity to interpret and make sense of what they experience in terms of those ideas. To help students generate conceptual understanding, however, they must also be guided in thinking about what they observe and in connecting their observations about instances to general principles and to what they already know. Student understanding of the learning goals increases when students (a) observe phenomena and representations that are explicitly targeted to the learning goals and are selected to address common student difficulties and (b) are actively engaged in interpreting the phenomena and representations in light of the learning goals and their own initial ideas. Consistent with these views, the *THSB* unit supports student learning by

- focusing on a coherent set of ideas and making the connections among those ideas clear,
- providing information about commonly held student ideas (both troublesome ideas and ideas teaching can build on) and questions to collect information about a particular group of students' ideas and to monitor their progress,
- including a variety of relevant and engaging phenomena and representations, and
- including activities that foster students' sense making, such as modeling tasks, to help students relate the phenomena and representations to the science ideas and to reconcile differences between their ideas and the science ideas.

With these design principles in mind, the AAAS and BSCS team developed the unit to address specific goals for student learning.

Goals for the Unit

The overarching goal of the *Toward High School Biology* unit is for students to apply ideas about what happens to atoms and molecules during chemical reactions to explain observable phenomena in nonliving systems and in the bodies of living organisms. Specifically, to grow and repair body structures, plants and animals build polymers through chemical reactions that link monomers and also produce water molecules. Animals get many of the monomers from breaking down other polymers in the foods they eat, whereas plants make monomers through other chemical reactions. During all these chemical reactions, atoms are rearranged and conserved; therefore, total mass is conserved.

By the end of the unit, students will be able to answer the following unit questions in terms of the rearrangement and conservation of atoms:

a. How do living things grow bigger?

b. Where does all the extra stuff come from as living things grow bigger?

The unit also addresses a number of important goals for science learning that are identified in the *Framework* (NRC 2012) and *NGSS* (NGSS Lead States 2013).

Alignment With National Science Education Framework and Standards

The *THSB* unit is designed to address core disciplinary ideas, science practices, and crosscutting concepts recommended as goals for learning in the National Research Council's *Framework* and in the *NGSS*. The core disciplinary ideas targeted in the unit are listed next and were drawn from the *Framework*, which provides a detailed description of them at each grade level. Science practices and crosscutting concepts are drawn from the *NGSS*, which clarifies them for each grade level in Appendix F and G, respectively. Parts of ideas and practices specifically targeted in the unit are shown in boldface, and a rationale for this selection is provided in the **Background Knowledge for Teachers** at the beginning of each chapter.

Disciplinary Core Ideas from the *Framework*

Physical Science, Structure and Properties of Matter (PS1.A, p. 108, Grade 8):

All substances are made from some 100 **different types of atoms, which combine with one another in various ways. Atoms form molecules that range in size from two to thousands of atoms. Pure substances are made from a single type of atom or molecule; each pure substance has characteristic** physical and chemical **properties** (for any bulk quantity under given conditions) **that can be used to identify it.**

Physical Science, Chemical Reactions (PS1.B, p. 111, Grade 8):

Many substances react chemically with other substances to form new substances with different properties. This change in properties results from the ways in which atoms from the original substances are combined and rearranged in the new substances. However, the total number of each type of atom is conserved in any chemical process, and thus mass does not change either. Some chemical reactions release energy; others store energy.

Life Science, Organization for Matter and Energy Flow in Organisms (LS1.C, p. 148, Grade 8):

Plants, algae, and many microorganisms use the energy from light to **make sugars (food) from carbon dioxide from the atmosphere and water through the process of photosynthesis, which also releases oxygen. These sugars can be used** immediately or stored **for growth** or later use. **Animals obtain food from eating plants or eating other animals. Within individual organisms, food moves through a series of chemical reactions in which it is broken down and rearranged to form new molecules, to support growth,** or to release energy. ...

Science Practices from the *NGSS*, Volume 2, Appendix F

Practice 2, Developing and Using Models (p. 53)

- Develop and/or **use multiple models to** predict and/or **describe phenomena.** (Grades 6–8)

Practice 4, Analyzing and Interpreting Data (p. 57)

- **Analyze and interpret data to provide evidence for phenomena.** (Grades 6–8)

Practice 6, Constructing Explanations and Designing Solutions (p. 61)

- **Apply scientific ideas, principles, and/or evidence to construct, revise, and/or use an explanation for real-world phenomena, examples, or events.** (Grades 6–8)

Practice 8, Obtaining, Evaluating, and Communicating Information (p. 65)

- **Critically read scientific texts adapted for classroom use to determine the central ideas and/or obtain scientific and/or technical information to describe patterns in and/or evidence about the natural** and designed **world(s).** (Grades 6–8)

Crosscutting Concepts from the *NGSS,* Volume 2, Appendix G

Energy and Matter: Flows, Cycles, and Conservation (p. 86)

- **Matter is conserved because atoms are conserved in physical and chemical processes. ...** (Grades 6–8)

These standards are addressed throughout the unit, and all three dimensions of science learning—core disciplinary ideas, science practices, and crosscutting concepts—are carefully integrated in each chapter. Students construct evidence-based explanations about phenomena. The evidence comes from direct observations students make during classroom activities and from patterns in data reported in the scientific research literature and summarized for students in the unit. Students use a variety of models to help them link the evidence to core ideas about atom rearrangement and conservation during chemical reactions in nonliving and living systems.

Performance Expectations from the *NGSS,* Volume 1

The unit contributes to the following middle and high school Performance Expectations:

- Analyze and interpret data on the properties of substances before and after the substances interact to determine if a chemical reaction has occurred. (MS-PS1-2)
- Develop and use a model to describe how the total number of atoms does not change in a chemical reaction and thus mass is conserved. (MS-PS1-5)
- Develop a model to describe how food is rearranged through chemical reactions forming new molecules that support growth and/or release energy as this matter moves through an organism. (MS-LS1-7)
- Construct a scientific explanation based on evidence for the role of photosynthesis in the cycling of matter and flow of energy into and out of organisms. (MS-LS1-6)
- Construct and revise an explanation based on evidence for how carbon, hydrogen, and oxygen from sugar molecules may combine with other elements to form amino acids and/or other large carbon-based molecules. (HS-LS1-6)

Structure of the Unit

The unit consists of 19 lessons organized into four chapters. Each chapter is organized around one or more of the core disciplinary ideas and crosscutting concepts that were listed. Students develop their understanding as they engage in the science practices of **analyzing and interpreting data, using models, constructing explanations, and reading scientific texts to obtain information and evidence.** Every lesson in the sequence supports students in developing the chapter learning goal, starting with the ideas and thinking students have as they enter the unit. Students have multiple exposures to the chapter learning goal throughout the chapter (and throughout the unit). Moreover, the learning goals in Chapters 1 and 2 are reinforced in Chapters 3 and 4, giving students the opportunity to examine examples of how the learning goals apply in both living and nonliving contexts. Many students require multiple examples to develop deep understanding of general principles; all students are not expected to grasp a new idea in the first lesson they encounter it.

Chapters 1 and 2 help students develop an understanding of chemical reactions by focusing on simple, nonliving systems. In Chapter 3 and 4, the central concepts developed in the first two chapters are applied to the more complex context of growth and repair in living organisms.

- Chapter 1 (Lessons 1.1–1.6) develops six science ideas related to the properties of substances and the production of new substances during chemical reactions and the molecular explanation for their production. Students first encounter the concepts in reactions involving only small molecules and apply them to reactions involving the formation of large macromolecules (polymers) and water molecules from small carbon-based molecules (monomers). Polymer formation is an important type of chemical reaction for understanding growth in living things, because plants and animals need to make polymers to build their body structures. Students develop explanations of phenomena using evidence from their observations and reasoning from science ideas and models.
- Chapter 2 (Lessons 2.1–2.3) develops four science ideas about mass conservation during chemical reactions and its molecular explanation. Students revisit the reactions they examined in Chapter 1, focusing on what happens to the mass when the reactions are carried out in closed versus open systems. They use models to visualize why even though atoms, and therefore total mass, remain the same (are conserved) during chemical reactions, the measured mass of a system can change if reactants or products enter or leave the system.
- Chapter 3 (Lessons 3.1–3.5) develops four science ideas as it applies the concepts of atom rearrangement and conservation to plant growth and repair: Through chemical reactions, plants build glucose monomers from CO_2 and H_2O they take in from the environment and link those glucose monomers to make the carbohydrate polymers (cellulose and starch) used to build body structures (which are made mainly of carbohydrate polymers). A plant's measured mass increases even though total mass is conserved. The increase in a plant's mass comes from mass in its environment (mainly from carbon dioxide gas): The more atoms that are added to a plant's body, the fewer atoms are in its environment.
- Chapter 4 (Lessons 4.1–4.5) develops three science ideas as it applies the concepts of atom rearrangement and conservation to animal growth and repair: Through chemical reactions, animals digest protein polymers in their food to amino acid monomers and use the monomers to build different protein polymers that make up their body structures (which are made mainly of protein polymers). An animal's measured mass increases even though total mass is conserved. The increase in an animal's mass comes from mass in its environment (mainly from the food it eats): The more atoms that are added to an animal's body, the fewer atoms are in its environment. In Lesson 4.5, students apply ideas about atom rearrangement and conservation in chemical reactions to the growth of mushrooms on a fallen dead tree.

Phenomena and Data

The unit is designed to provide students with opportunities to both experience phenomena directly and to examine data collected by scientists that can serve as evidence for the science ideas or for their explanatory power. The phenomena and data include numerous examples from both living and nonliving contexts. The unit includes tasks and questions that help students (1) explain observable phenomena and data using accepted science ideas and models of underlying molecular events and (2) apply science ideas developed in nonliving contexts to living contexts, thus making visible a coherent science content storyline for the molecular basis of growth and repair in living organisms.

Representations

The unit uses a variety of representations of atoms, molecules, and chemical reactions. By the end of the unit, students should be able to translate fluently among these representations and use them to communicate what happens during chemical reactions.

LEGO models are used to represent atoms and molecules. In LEGO models, an atom is represented by a LEGO brick, and different-colored LEGO bricks represent different kinds of atoms. Colors follow scientific conventions. A molecule is represented by attaching LEGO bricks together in specific ways. When representing a certain molecule, the LEGO bricks are always attached to the same points, facing the same direction, and with the same number of clicks. LEGO models do not accurately represent molecular shape or show bonds between atoms. However, they can be used to represent atom rearrangement and conservation and are easy for students to manipulate when modeling the changes substances undergo during chemical reactions. While we use "rules" for assembling LEGO models, it is not necessary that students know these rules. The point to make when using LEGO models is that bricks of starting models are rearranged into different models (Chapter 1), and this rearrangement, without the need to add or eliminate LEGO bricks, represents the conservation of atoms during chemical reactions, which accounts for conservation of mass (Chapter 2).

Molecular models include two-dimensional **space-filling models** and three-dimensional **ball-and-stick** models, photographs, or drawings of those models. In these models, each atom is represented as a sphere. Spheres of different colors and sizes represent different types of atoms. Colors follow scientific conventions. Color images of models are available on the book's Extras page at *www.nsta.org/towardhsbio*. You might want to remind students that atoms do not really have these colors—this is just a tool for identifying different kinds of atoms. A molecule is represented by two or more spheres connected together in a specific arrangement, without sticks (space-filling models) or with sticks (ball-and-stick models). These models show us the types of atoms that make up a molecule and how many of each type. They also show how the atoms are arranged. This gives us an idea of how molecules are shaped and how much space they occupy (how small or large they are) in comparison with other molecules. When using the ball-and-stick models, you should purposely limit the discussion about sticks to "a way to represent connections between atoms."

Chemical names, symbols, and **formulas** are used to represent atoms and molecules. Each type of atom is represented with a **symbol** that is an abbreviation of its English or Latin name. A symbol is usually one or two letters, with the first letter always capitalized. A molecule is represented with a **formula** that shows the symbol of each atom making up the molecule and follows the symbol with a subscript indicating the number of atoms of that type that are present. If no number follows a symbol, it implies that the molecule has only one of that type of atom. Formulas do not show how the atoms making up a molecule are connected.

Chemical reactions are represented by **model-building activities, equations,** and **chemical reaction mats**. Students manipulate LEGO or ball-and-stick models to convert molecules of reactants into molecules of products, and weigh models in sealed versus open baggies to represent mass conservation and changes in measured mass. They use word equations with chemical names (e.g., iron in steel wool and oxygen in the air reacts to form iron oxide) and equations with chemical symbols and formulas ($4\ Fe + 3O_2 \rightarrow 2\ Fe_2O_3$). And they look at chemical reaction mats showing words, formulas, and photos of molecular models of reactants on one side (green) and flip the mats to the other (yellow) side to see the words, formulas, and photos of molecular models of products. The act of flipping the mat over represents the "react to form" arrow in the equation.

Toward High School Biology (*THSB*) Content Storyline

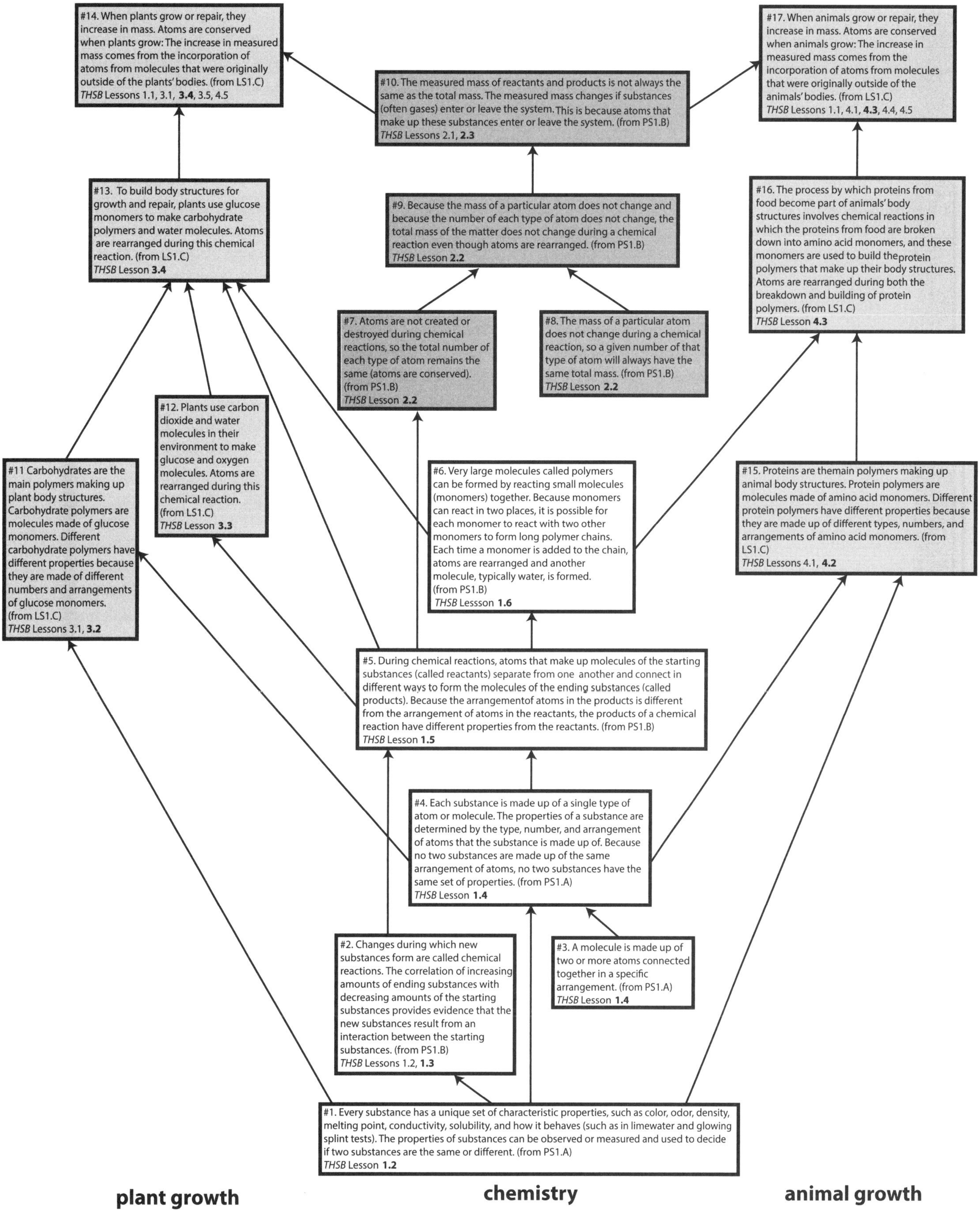

Unique Aspects of the Unit

Understanding some of the thinking that went into the design of *Toward High School Biology* and what the unit is intended to do can help you support students in achieving the unit's learning goals. This section briefly describes the development team's rationale for decisions about how the core science ideas, practices, and crosscutting concepts would be integrated in the unit, what vocabulary students would be expected to know and use, and the extent to which the unit contributes to students' understanding of the nature of science.

Integrating Physical and Life Science Ideas About Chemical Reactions

In many school districts, life science is offered in the seventh grade and physical science in the eighth grade, so students typically encounter photosynthesis and cellular respiration before they are taught what chemical reactions are. Unless physical science and life science teachers coordinate instructional planning with coherence in mind, students will emerge from both courses with fragmented knowledge of the natural world. The THSB unit takes a different approach, treating ideas about chemical reactions in both living and nonliving contexts together and taught by the same teacher, so that connections among the ideas can be readily made. This integration of physical and life science content is consistent with recommendations of the *NGSS* in general and in particular with its crosscutting concepts related to matter and energy. Content support is provided in both the Teacher Edition and in professional development so that teachers of either life or physical science can become confident teaching all aspects of the unit.

Matter

The unit targets all of the core ideas about changes in matter that the *NGSS* recommend for middle school students. This includes ideas about chemical reactions in physical and life science at both the substance and at the atomic/molecular levels. These are the ideas needed to understand how new substances can be produced for growth and repair of living organisms from starting substances that appear quite different and why mass increases accompanying the growth of animals and plants do not violate conservation principles. The unit does not target more sophisticated high school ideas about atomic structure and chemical bonding because the development team felt that adding those ideas would needlessly diminish the unit's coherence.

Energy

In addressing *NGSS* core ideas and crosscutting concepts related to matter and energy, we needed to make decisions about what an eight-week unit that also targeted science practices could realistically accomplish. As a result, the unit focuses mainly on ideas dealing with matter and does not target energy ideas explicitly. There were both practical and conceptual reasons for this decision. Assessment data from AAAS's prior work indicated that students lacked mental models of both matter transformation and energy transformation, so we did not think eight weeks would be sufficient time to tackle both. Because matter transformation could easily be made concrete with physical models, we opted to start there. During pilot testing of an early version of the unit, we experimented with including a single simplified idea about energy ("atoms don't turn into energy and energy doesn't turn into atoms") to address a common misconception. We did find a decrease in the prevalence of the matter-from-energy misconception when students used that initial version of the unit. However, we observed that (a) only a few students actually invoked the idea when challenging their classmates' explanations, so we questioned its utility in the context of the unit and (b) only a few students were able to use an atomic/molecular model to explain chemical reactions in open systems, so we knew there was still much more work to be done in this area. With only eight weeks of instructional time available to us, we decided to focus subsequent versions of the unit entirely on matter ideas.

Cells

For several reasons, the *THSB* unit "black boxes" events related to biological growth and repair at the cellular level. Because the main goal of the unit is to help students understand the role of chemical reactions in biological growth and repair, we wanted to keep the focus on the properties of the biomaterials (protein polymers, carbohydrate polymers, and composites of them), which provides students with a more powerful—and demonstrable—explanation for the properties of body structures (e.g., muscle tissue or plant stems) than the fact that they are made up of cells. We also wanted to preserve the content storyline, which connects reactions in nonliving systems (e.g., polymerization of nylon) to similar reactions in living organisms (e.g., polymerization of proteins and carbohydrates). Focusing on the role of cells in biosynthesis would have distracted from that central storyline. Moreover, we thought that a consistent focus on linking mass/matter/materials/atoms to growth would be more likely to extinguish the common student misconception that cell division alone accounts for growth, which tends to be reinforced by existing curriculum materials.

Science Practices

Engaging students in scientific practices is central to the design of the *THSB* unit. Students use a variety of molecular models to make sense of atom rearrangement and conservation in chemical reactions; they analyze data drawn from their reading of simplified accounts of scientific research; they use science ideas and evidence from firsthand observations and data to support explanations of real-world phenomena. Not only do we want students to understand and appreciate the role these practices play in constructing scientific knowledge and evaluating explanations, we also view them as highly motivating and engaging instructional strategies. Designing learning experiences to promote the scientific practices of modeling, data analysis, and communication was a fairly straightforward task, and we were guided by the recommendations in the *Framework* and in the *NGSS*. Considerable effort went into designing and testing scaffolds to support explanation writing as described below.

Constructing Explanations

The *THSB* unit engages students in constructing scientific explanations of phenomena. Because of the complexity of this practice and the reported difficulty students have with constructing explanations, *THSB* provides support consistent with the cognitive apprenticeship model (Brown, Collins, and Duguid 1989; Collins, Brown, and Newman 1989). In the beginning of the unit, students are provided with an example of a valid explanation, and the essential elements of an explanation—claim, relevant science ideas, and evidence that will be linked to the claim using science ideas—are defined. The early lessons also establish Explanation Quality Criteria that students can use in judging an explanation's quality and introduce a template to remind them about the essential elements and help them organize their thinking and writing. After students experience modeling activities, a new element is added to the explanation template—models/modeling as tools for thinking about how something might work—and from then on students are expected to include models in their reasoning. The cognitive apprenticeship cycle is then repeated using the full explanation template. As the unit proceeds, students are reminded to use ideas about atoms in their explanations and to be sure that their explanations meet the Explanation Quality Criteria. The explanation template is faded in the Student Edition but is available in the Teacher Edition if the teacher feels the students still need it. At the end of the unit, the scaffolding is almost completely removed with the exception of a reminder that students can use science ideas from anywhere in the unit in their explanation.

The Teacher Edition and web-based teacher resources support teachers in helping students construct evidence-based explanations. The Teacher Edition provides ideal responses for each explanation task and for each element of the explanation template. Online teacher resources provide rubrics for explanation tasks recommended for use as embedded assessments. Rubrics list the specific elements of a complete explanation of each phenomenon, including claim, science ideas to be invoked, evidence to be cited that is consistent with those science ideas, reasoning that uses the science ideas, and, when appropriate, reasoning that uses ideas from modeling activities.

Nature of Science

Although the THSB unit was not designed to make ideas about the nature of science explicit, there are many opportunities throughout the unit for students to experience how science works. For instance, students consistently use models to think about and visualize atom rearrangement and conservation during a variety of chemical reactions and they analyze data and develop evidence-based explanations in ways that are consistent with what is expected of scientists. Students are repeatedly asked to list examples of phenomena they have observed in class that are consistent with presented science ideas and are told that the reason they can legitimately use the science ideas to support explanations of similar phenomena is because the science ideas are consistent with a wide range of observations and data—not just the small number they examined in class. Finally, students read simplified versions of scientific papers and examine how scientists use data in supporting their conclusions.

Nonetheless, the *NGSS* notes that while integrating science practices, core ideas, and crosscutting concepts "sets the stage for teaching and learning about the nature of science" (NGSS Lead States 2013, p. 97, Appendix H), engaging in such activities is not sufficient. Students would also need to reflect on their model-building, data analysis, and explanation activities and to compare their work to the work of scientists, which would likely involve examination of historical case studies. The THSB unit uses Van Helmont's willow tree experiment as a case study. However, given the time that these additional activities would require and the risk they might pose to the coherence of the content storyline, we reluctantly decided not to focus more instructional time on nature of science learning goals.

Vocabulary

We have intentionally limited the number of technical terms to those that students need to communicate about their experiences. As a result, the unit avoids several terms typically used in eighth grade classrooms (e.g., element, compound, ion) while using other terms that are less typical (e.g., monomer, polymer). The terms *element, compound,* and *ion* are not needed for students to communicate their ideas about atom rearrangement and conservation, although students do need to correctly use the terms *atom* and *molecule* to do so. The terms *monomer* and *polymer* are used instead of alternatives like *small molecule* and *large molecule* because they convey more precisely the common features of carbohydrates and proteins and how they can be assembled from or disassembled to a small number of building blocks. We also leave out some details that are more likely to confuse than to help students develop and communicate their ideas. For example, solutions are referred to as liquids and ionic compounds are referred to as molecules.

The unit expects students to understand and correctly use the following terms in **bold** related to matter:

> **Atoms** and **molecules.** While students may have already been taught definitions of these terms, many do not consistently distinguish them in their talking or writing or in their use of models. It is essential that students know that atoms make up molecules, not the other way around. The unit expects students to know that an individual atom is represented by a ball or LEGO brick, that different colored balls or bricks represent different types of atoms, and that two or more balls or bricks linked together represent molecules.

Monomers and **polymers.** Rather than relying on vague descriptions like *small and large molecules,* which provide no information about how the small and large molecules are related, we opted to introduce the terms *monomer* and *polymer*. Using these terms in the context of chemical reactions involved in nylon formation, protein formation, and cellulose formation highlights their similarities (all involve formation of polymers and water molecules from monomers) and eliminates the need for students to know the names of most of the specific monomers involved (e.g., hexamethylenediamine, adipic acid, amino acids). However, students are expected to know that **glucose** is the monomer used to build carbohydrate polymers, such as cellulose and starch. The term *glucose* is singled out so that students can distinguish its formation (during **photosynthesis**) from its use to make polymers needed to build plant body structures for growth. Because the term *carbohydrates* refers to both polymers and monomers, accuracy requires that we use the term *carbohydrate polymers*. Students are expected to know that proteins are **polymers**.

Substances, materials, and **body structures.** The unit distinguishes pure substances (such as water, glucose, cellulose, and ovalbumin that are made of a single type of molecule) from materials (such as egg white that are made up of several types of molecules) from body structures (such as muscles and plant stems that are made from a variety of materials). The unit does not digress to discuss the relationship between molecular structure and properties of substances or materials made from them, which requires an understanding of atomic structure and chemical bonding.

Chemical reactions. Students are expected to know that chemical reactions produce **products** from **reactants**, that products have different **characteristic properties** from the reactants because they are made up of different molecules, but that the molecules making up the products are made from the atoms of the reactants. Students are expected to distinguish chemical reactions from changes in matter that do not involve chemical reactions based on comparing the molecules of the starting substances to the molecules of the ending substances. (Irreversibility is not a criterion for distinguishing chemical reactions from other changes—this is a common misconception). Students are expected to know that **photosynthesis** and **digestion** and the synthesis of proteins and complex carbohydrates involve chemical reactions.

Mass conservation and **measured mass.** Students are expected to know that while **total mass** of the reactants and products does not change during a chemical reaction (total mass is **conserved**), the **measured mass** may change if one or more substances enter or leave the **system**. The term *measured mass* is introduced to help students understand why conservation principles aren't violated during biological growth; that is, the added mass has to have come from somewhere outside the system being measured (the plant or animal's body). The unit engages students in weighing models of reactants and products of reactions in both closed and open systems to help them visualize that both mass and atoms (though not necessarily molecules) are conserved during chemical reactions.

The unit expects students to understand and correctly use the following terms in **bold** related to constructing and evaluating explanations:

Evidence. Students are expected to know that **data** from observations and measurements can be used as evidence for an explanation if the data are relevant to the answer and can be confirmed by others. Students are also expected to know that a **model** can sometimes be used to get ideas about how the thing being modeled actually works, but there is no guarantee that these ideas are correct if they are based on the model alone. Consequently, ideas from models do not provide evidence to support claims but should be consistent with claims.

Explanation. As noted in the *NGSS*, "the goal of science is to construct explanations for the causes of phenomena" (NGSS Lead States, 2013, p. 60, Appendix F). Throughout the unit, students are expected to explain phenomena in terms of underlying molecular events. Students are also expected to evaluate explanations of others. Students learn that an explanation must be consistent with all the evidence, relevant science ideas, and models.

Student Edition Conventions

Each chapter of the *Toward High School Biology* unit consists of carefully sequenced lessons and activities designed to (1) draw upon students' prior knowledge and experiences relevant to classroom activities; (2) support students as they investigate and make sense of phenomena and models; (3) guide students in developing, analyzing, and critiquing explanations (e.g., of a hypothetical student, of their peers, and of those of the scientific community) in light of their experiences; (4) provide opportunities for students to apply or extend science ideas and practices to new phenomena; and (5) help students synthesize their ideas and reflect on changes in their thinking.

Section Headings and Purposes

Each lesson is designed with particular features that are denoted by their section headings.

What do we know and what are we trying to find out?

This introductory section situates the lesson in the content storyline by making links between science ideas of previous lesson(s) and the key question of the current lesson. This section does not "give away the answers" but provides students with some sense of what they will be working toward understanding in the lesson.

Key Question

Each lesson begins with and returns to a key question that the lesson is designed to answer. The key question aligns with the lesson's main learning goal and frames the students' inquiry. The key question is posed at the beginning of the lesson to give each student a chance to express his or her initial ideas and to give teachers a sense of the range of ideas students hold. At the end of the lesson, the question is posed again and can be used to monitor students' progress.

Activity

Each lesson includes activities designed to engage students with phenomena and representations relevant to the learning goals. Some phenomena, particularly those in nonliving systems, can be observed directly and others, particularly those occurring in the bodies of living organisms, require inferences from data. Activities also engage students in modeling invisible aspects of the phenomena, particularly atom rearrangement and conservation. Each activity includes questions to focus and guide students in observing and interpreting the phenomena. Activities are structured to encourage discussion by having students work either with a partner or a small group.

Some activities consist of short readings and questions to (1) help students make sense of ideas presented in text, (2) encourage students to ponder what they have just read in manageable chunks, and (3) help students use and apply the ideas they have read in the text.

Other activities provide opportunities for students to learn to construct explanations, evaluate sample explanations, and develop their own explanations for phenomena they encounter in the unit and in their everyday lives.

Participating in a whole-class discussion at the end of an activity can help students reach consensus on their observations of phenomena, data, and models and on the ideas that emerge from their interpretations of them.

Science Ideas

At critical points in the unit, particularly after students have developed ideas based on phenomena, data, or models, students are asked to generalize across their experiences. Only then are relevant science ideas introduced to students as generalizations about how the world works based on a wider range of observations and data. (Each science idea, followed by a list of the lessons in which it plays a role, is shown in a box on the Content Storyline map on p. xi. Lesson numbers in bold indicate where a science idea is first introduced.) Students have opportunities to compare the ideas they are developing with established science ideas and find examples from their work that support the science ideas.

Science ideas are used in writing explanations, so they will often precede an activity in which explanations are evaluated or developed. Or, they may precede a Pulling It Together section in which students are expected to construct an explanation.

Pulling It Together

These questions provide opportunities for students to individually (1) revisit and answer the lesson key question or related questions to summarize their current understanding or new learning, (2) use and apply the ideas they are developing to a new context or phenomenon, and (3) begin to link the ideas to the next lesson(s). The linking question elicits student ideas and predictions about the key question of the next lesson or about how an idea developed in nonliving contexts might apply to growth in living things.

References

Brown, J., A. Collins, and P. Duguid. 1989. Situated cognition and the culture of learning. *Educational Researcher* 18 (4): 32–42.

Collins, A., J. S. Brown, and S. Newman. 1989. Cognitive apprenticeship: Teaching the crafts of reading, writing, and mathematics. In *Knowing, learning, and instruction: Essays in honor of Robert Glaser,* ed. L. B. Resnick, 453–494. Hillsdale, NJ: Erlbaum.

NGSS Lead States. 2013. *Next Generation Science Standards: For states, by states.* Washington, DC: National Academies Press. *www.nextgenscience.org/next-generation-science-standards.*

NRC (National Research Council). 2012. *A framework for K–12 science education: Practices, crosscutting concepts, and core ideas.* Washington, DC: National Academies Press.

The following page has a blank explanation chart. Initially all explanation tasks have this writing scaffold. As the unit progresses, the scaffold is removed. If students need additional scaffolding later in the unit, teachers may copy the blank chart for use with their students.

Question	
Claim	
Science Ideas	
Evidence	
Models	

Explanation:

Toward High School Biology

Chapter 1

Teacher Edition

Chapter 1 Overview

Students are familiar with many everyday changes that matter undergoes, in both living and nonliving things. They observe substances changing naturally and they participate in changing other substances themselves. Students may also be quite familiar with atoms and molecules. Too often, however, students do not share scientists' ideas about an "atom rearrangement model" when it comes to explaining the changes they witness on a daily basis. Chapter 1 helps students develop the concepts that **different substances have different characteristic properties because they are made up of different arrangements of atoms** and that **new substances form during chemical reactions because atoms making up molecules of the reactants rearrange to form molecules of the products.**

Lesson 1.1 provides students with an opportunity to observe changes in matter that makes up both living and nonliving things. Students are asked to try to identify similarities in these changes (e.g., new materials form or "grow," the matter of new materials comes from other materials, and the weight of the new matter increases as it "grows"), but they will have numerous opportunities throughout the unit to revisit and revise their ideas as they learn more concretely how these changes are similar.

Lesson 1.2 transitions students from changes in complex mixtures of substances or materials that make up living things to three examples of pure substances reacting in simple systems that will be used to illustrate and help students generalize important concepts about chemical reactions. Students gather firsthand observations of samples of reactants, chemical reactions, and samples of the products and examine data to form the generalization that different substances have different sets of characteristic properties that can be used to tell one substance from another. Students are introduced to criteria for judging the quality of explanations and use the criteria to evaluate and construct evidence-based explanations to support claims that new substances form during these changes.

Lesson 1.3 revisits the same three chemical reactions introduced in Lesson 1.2 but focuses students on observing changes in the amounts of starting and ending substances as the reaction proceeds, specifically that as the amounts of the ending substances forming increase, the amounts of starting substances decrease. From their observations, students should infer that the ending substances probably come from the starting substances.

Lesson 1.4 investigates why, for each change, the ending substances have different properties from the starting substances. Students examine models to visualize differences in the type of atoms, number of each type, and how the atoms are arranged in different substances and generalize that whenever substances have different properties they are made up of different arrangements of atoms. Students also begin making sense of the three reactions in terms of the types of atoms and molecules that make up the ending substances as compared with starting substances from which they are made.

Lesson 1.5 is a structured modeling activity with LEGO bricks that guides students in developing the concept that atoms making up molecules of the reactants are rearranged to form the molecules of products during chemical reactions. Students are introduced to the role that models can play in supporting explanations. The activity focuses on the iron rusting reaction and the reaction between baking soda and vinegar and to having students expand their explanations to include what is happening to the atoms and molecules to account for the production of ending substances with different properties.

Lesson 1.6 applies the overarching concepts about chemical reactions to the formation of very large molecules (polymers) from small molecules (monomers). Students examine characteristic properties of the monomers and the polymer and use ball-and-stick models to model the atom rearrangement that occurs when nylon and water molecules are formed from hexamethylenediamine and adipic acid monomers. Typical of polymerization reactions that students will encounter in Chapters 3 and 4, nylon formation involves the rearrangement of only a few atoms as the monomers are linked and water molecules are formed.

Chapter 1 Background Knowledge for Teachers

The Three Chemical Reactions for Chapters 1 and 2

Three chemical reactions are the backbone of the first two chapters: the reaction between baking soda and vinegar, between iron and oxygen, and between hexamethylenediamine and adipic acid. Students investigate these chemical reactions in depth—from both macroscopic and microscopic perspectives. Chapter 1 focuses on how the rearrangement of the atoms making up the molecules of the starting substances can explain the production of substances with different properties: atom rearrangement produces different molecules, giving the products different properties from the starting substances. Chapter 2 focuses on helping students understand that if atoms are merely rearranged, the type and number of atoms does not change (atoms are conserved); therefore, the total mass does not change (total mass is conserved). It also helps students reconcile conservation ideas with changes in mass of a system: While atoms are not created or destroyed, the mass being measured may change if atoms enter or leave the system. The Chapter 2 Background Knowledge document contains discussion of mass for each chemical reaction.

Baking Soda and Vinegar

The reaction between vinegar (acetic acid, $HC_2H_3O_2$) and baking soda (sodium bicarbonate, $NaHCO_3$) actually involves two steps—an acid-base reaction followed by a decomposition reaction. When vinegar and baking soda are mixed, hydrogen ions (H^+) from the acetic acid in vinegar react with the bicarbonate ions (HCO_3^-) from the baking soda to form a new chemical called carbonic acid (H_2CO_3):

$$H^+ (aq) + HCO_3^- (aq) \rightarrow H_2CO_3 (aq)$$

The carbonic acid formed then decomposes into carbon dioxide gas (CO_2) and water (H_2O). It is carbon dioxide gas that you see bubbling as soon as you mix baking soda and vinegar together.

$$H_2CO_3 (aq) \rightarrow H_2O (l) + CO_2 (g)$$

The following chemical equation describes the net reaction between vinegar and baking soda that forms soluble sodium acetate ($NaC_2H_3O_2$), water, and carbon dioxide gas:

$$\mathbf{HC_2H_3O_2 (aq) + NaHCO_3 (s) \rightarrow NaC_2H_3O_2 (aq) + H_2O (l) + CO_2 (g)}$$

Intended observations: Students are expected to see gas bubbles forming, hear fizzing, and feel increased gas pressure in the container when the clear, colorless liquid and white powder are mixed. Students may also observe a diminished odor if enough of the vinegar is consumed during the reaction.

Rationale for including this reaction: This acid-base neutralization reaction provides an example of non-gaseous reactants yielding gaseous products. It is used in Chapter 2 to provide evidence that gases have mass, because students can observe that the mass of the container decreases when gas is allowed to escape.

Steel Wool (Iron) and Air (Oxygen)

Steel wool is made mostly of the element iron (Fe). To prevent the iron from rusting, steel wool is made with an oil-based coating, which can be removed with vinegar or heating. The rusting of iron is a complex electrochemical process that begins with the transfer of electrons from iron to oxygen and involves multiple oxidation/reduction and acid-base reactions. The products of rust formation depend on the amount of oxygen and water present (i.e., FeO in oxygen-depleted conditions, Fe_2O_3 in oxygen-rich conditions). The following chemical equation is commonly used to represent the process overall, leaving out various intermediate products that occur temporarily along the way:

$$4\,Fe\,(s) + 3\,O_2\,(g) \rightarrow 2\,Fe_2O_3\,(s)$$

The reaction of Fe and O_2 is relatively slow but is speeded up by water and ions, which is why cars rust when in contact with salted roads. Given sufficient time and oxygen, any iron sample will eventually covert completely to rust. The Fe_2O_3 rust product is a delicate and brittle reddish-brown solid that breaks apart with the slightest touch. This change in texture often leads to misconceptions that rusting causes a decrease in mass (which is what happens when rust falls off), as opposed to the increase in mass caused by chemically combining iron and oxygen.

Intended observations: Students should observe rust products that form on steel wool. By carrying out the reaction in a flask sealed with a balloon, they will also have evidence that a gas in the flask is involved in the reaction (i.e., the balloon is pushed or inverted into the flask as the oxygen in the air is used, causing the pressure inside the flask to decrease). Trace amounts of water speed up the reaction but is regenerated at the end, so the amount doesn't change from start to finish. Because the rusting reaction releases heat, the water may vaporize during the reaction and then condense once the reaction is complete. Students may think that water is a product, but the water was there from the start. Point this out to focus students' attention on the overall reaction.

Rationale for including this reaction: In Chapter 1, rusting provides an example of a gaseous reactant yielding a solid product. This is analogous to photosynthesis reactions that "fix" gaseous carbon dioxide from the air to form solid glucose for building plant materials.

Hexamethylenediamine and Adipic Acid

Nylon is formed at the interface of two non-mixing liquids—a diamine, hexamethylenediamine ($NH_2(CH_2)_6NH_2$) and a diacid, adipic acid ($HOOC(CH_2)_4COOH$). When the film is drawn or spooled from the interface of the two liquids, a thread of nylon polymer forms. This thread is composed of huge numbers of polymer molecules that are cross-linked together.

The reaction between one hexamethylenediamine monomer and one adipic acid monomer is shown below:

$$C_6H_{16}N_2 + C_6H_{10}O_4 \rightarrow C_{12}H_{24}O_3N_2 + H_2O$$

Many chemical reactions that form polymers, including nylon and biological polymers, also produce water. Often this type of chemical reaction occurs between two or more different carbon-based small molecules (monomers) that will alternate in the polymer chain. Each monomer has specific functional groups (arrangements of atoms) at both ends of the molecule—either a carboxylic acid group (COOH) or an alcohol group (OH) and/or an amine group (NH2) that allow it to link to other monomers of the growing polymer chain.

During the chemical reaction, an OH from one monomer reacts with an H from another monomer to form a water molecule and a longer polymer as shown below.

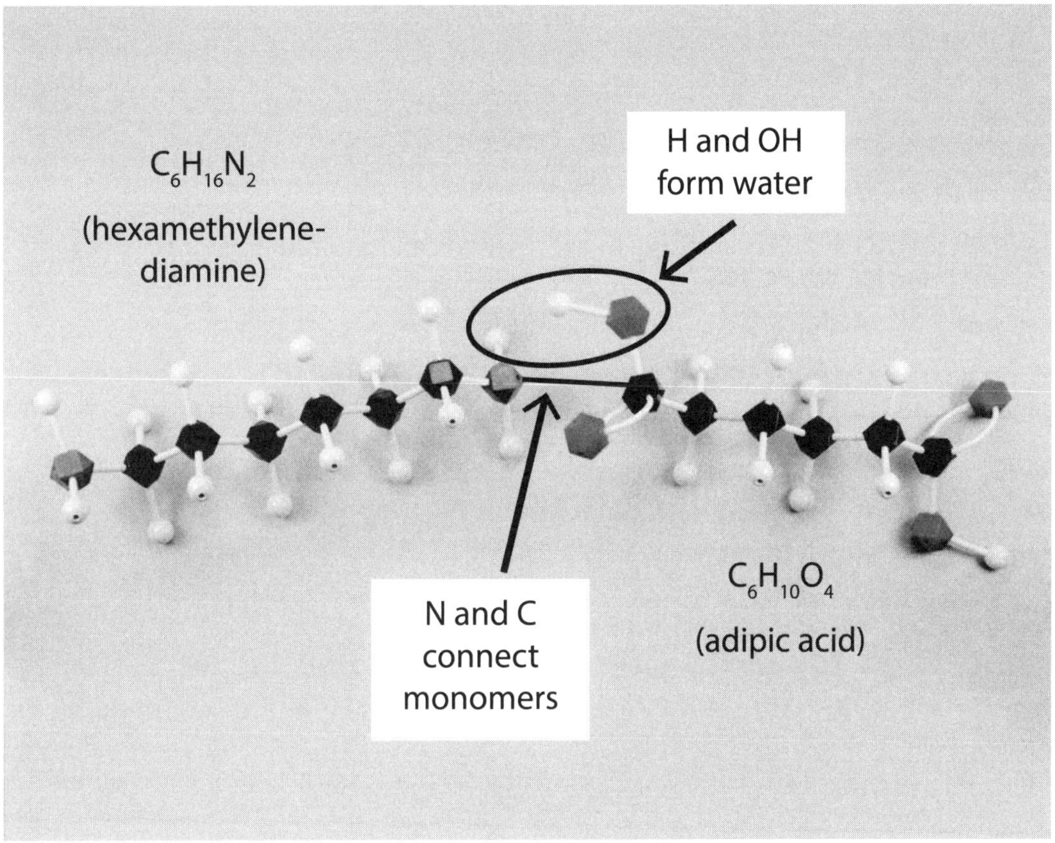

Because each end of a monomer can react, long polymers can form. As each monomer is added to a growing polymer chain, a water molecule is produced.

Intended observations: Students should observe the formation of a white fibrous solid (nylon) from two clear, colorless liquids. Students should also observe that as the nylon is formed the amount of the liquids in the beaker decreases.

Rationale for including this reaction: The formation of nylon from hexamethylenediamine and adipic acid is an example of the formation of polymers from monomers through chemical reactions in which only a limited number of atoms rearrange. Polymer formation is key to understanding the formation of protein- and cellulose-based body structures through the polymerization of amino acids and glucose, respectively.

Chapter 1 Prerequisite Knowledge for Students

Lessons 1 and 2 provide entry points for most students, despite their prior knowledge. However, the chapter assumes that students know that all matter is made up of atoms. (The lessons will develop the idea that atoms can link together to form a variety of different molecules.)

Although the unit does not emphasize physical changes, students will learn that because atoms do not rearrange to form new molecules during physical changes, such changes are not chemical reactions.

Chapter 1 Commonly Held Student Ideas and Misconceptions

Researchers have documented many incorrect ideas that students commonly have about chemical reactions. This document provides a brief summary of most of the more common incorrect ideas, but it is not exhaustive. Listen for these ideas in your class, but also for other variants that your particular students may have. Actual student responses will usually be shorter and less precise than those listed here. The descriptions here are meant to fully explain the nature of the incorrect idea and to highlight what is unsaid as much as what is said.

Some students have trouble distinguishing between atoms and molecules, and are often imprecise in the language, referring generally to "atoms and molecules." Students may say:

- *Atoms are "in" matter (rather than "make up" matter).*
- *Atoms and molecules are basically the same—they are all just very small particles.*
- *Molecules make up atoms.*

Some students have trouble distinguishing one substance from another based on their characteristic properties. Students may say:

- *If two substances share one characteristic property, then they must be the same substance.*
- *If most of the listed characteristic properties are the same, the substances must be the same.*
- *Substances can change their characteristic properties but still remain the same substance.*
- *The properties of a substance are determined (only) by the types of atoms that make up the substance (with no reference to the number of atoms or the way in which the atoms are arranged into molecules.)*
- *Two substances that are made up of molecules that have the same number of the same types of atoms are the same substance (regardless of how the atoms are arranged.)*

Many students use incorrect criteria for deciding whether or not a chemical reaction has occurred. Students may say:

- *Whenever matter changes, it is a chemical reaction.*
- *Chemical reactions happen whenever you combine two substances/chemicals.*
- *Chemical reactions always have two reactants (A + B → C OR A + B → C + D).*
- *Chemical reactions are irreversible changes. If you can "change it back," it is not a chemical reaction.*
- *If there are "chemicals" (judged based on their chemical names) involved, it must be a chemical reaction.*
- *A change of state is a chemical reaction.*
- *Whenever a gas forms, it is a chemical reaction.*
- *If heat is involved, it is a chemical reaction.*
- *If it has explosions, smoke, or color changes, it is a chemical reaction. Less visible changes are not chemical reactions.*

Many students may also have difficulty understanding how the products relate to the reactants of a chemical reaction. Students may say:

- *The products do not form from the reactants. The reactants disappear and the products appear.*

- *The products are just mixtures of the reactants, not totally new substances.*
- *The products are the same substances as the reactants, just with different properties.*

Students also commonly have misunderstandings about what happens to the atoms and molecules during a chemical reaction. Students may say:

- *The atoms of the reactants are transformed into other types of atoms to form the products. For example, an oxygen atom could become a carbon atom.*
- *All the atoms must rearrange (none can maintain attachments) for a chemical reaction to occur.*
- *During a chemical reaction, the molecules of the reactants stay the same and just stick together at their ends to form one big molecule, which is the product.*
- *The molecules of any one product must include every type of atom from the reactants.*

Lesson Guide
Chapter 1, Lesson 1.1
Changes in Living and Nonliving Things

Focus: This lesson focuses students' attention on changes that occur in living and nonliving things, encouraging them to consider what these changes have in common. For example, students might notice that in both nonliving and living systems new materials form or "grow," the "stuff" of new materials comes from other materials, and the weight of the new stuff increases as it "grows." Students will have numerous opportunities throughout the unit to revisit and revise their ideas as they learn more concretely how these changes are similar.

Key Question: How are changes in the matter that makes up living and nonliving things similar?

Target Idea(s) Addressed

Note: The lesson serves to promote and elicit students' thinking about similarities in the changes that occur in living and nonliving things. There are no target ideas or science practices specifically addressed in this lesson, but the phenomena related to changes in matter will contribute to students' understanding of how new substances form during chemical reactions.

Materials

Activity 1: Videos: *Puppy Grows Into a Dog, Corn Growth, Spider Spinning Silk, Hexamethylenediamine and Adipic Acid*

Phenomena, Data, or Models	Intended Observations	Purpose	Rationale and Notes
Activity 1 Time-lapse photography of dog growth and corn growth. Video of a spider spinning a web and video of nylon formation.	Changes occur in living and nonliving things. During these changes the following happens: • New materials are made (e.g., fur, skin, bones, and muscle) as a puppy grows. • New materials are made (e.g., leaves, stem, fruit) as a corn plant grows. • New materials are made from other materials (e.g., nylon thread is made from the liquids in the beaker that are used up and spider silk is made from substances in the spider's body).	To focus students' attention on similarities in the changes that occur in living and nonliving things.	Identifying similarities in the changes that occur in living and nonliving things may suggest to students that changes occurring in both living and nonliving things are guided by the same "fundamental principles" of science. Some students may struggle to make the connections we identified in the intended observations. Do not make the connections for them at this point. They will have numerous opportunities throughout the unit to revisit and revise their ideas as they learn more concretely how these seemingly unrelated changes are connected through fundamental chemical principles.

Lesson 1.1—Changes in Living and Nonliving Things

What do we know and what are we trying to find out?

Everything in the world—both things that are alive and things that aren't—is made of "stuff" that is changing all of the time. Scientists call this "stuff" matter. For example, a bicycle is matter, and when it is left outside, it becomes covered with rust, which is also matter. Matter changes when a baker makes a cake from flour, sugar, eggs, baking powder, and butter. Students mix vinegar with baking soda and change these ingredients into a foaming "volcano." Wood burns in a campfire and turns to ashes. Scientists in the lab combine different substances to create new products.

Living things are also made of matter, and most of them—including you—change as they grow and repair their bodies (see the color version of Figure 1.1 that your teacher will project). A small puppy grows into a big dog. A tiny seed sprouts a seedling that then grows into a giant sequoia tree. The skin on your finger heals around a cut. A lizard regrows a tail that it lost to a predator.

Figure 1.1. Changes in Matter

You can probably think of many more examples of how matter changes. But how do these changes happen? Do living and nonliving things change in the same way?

To answer these questions, we need to understand more about the "stuff" that makes up living and nonliving things and how it changes. In this lesson, you will observe some of these changes and begin thinking about the Key Question (there is no need to respond in writing now).

Key Question: How are changes in the matter that makes up living and nonliving things similar?

Teacher Talk and Actions

What do we know and what are we trying to find out? (3 min)

Whole class:

Read text aloud and examine with the class the images shown under "What do we know and what are we trying to find out?"

Key Question (1 min)

Introduce the Key Question: **How are changes in the matter that makes up living and nonliving things similar?**

Usually, you will use the Key Question to elicit and probe student ideas.

In Lesson 1.1, students do not need to respond to it before experiencing the activity. The activity is intended to provide students with concrete examples of changes in living and nonliving things that they can compare. The lesson culminates in students answering the Key Question based on their observations and comparisons of the changes.

Activity 1: Observing Changes in Living and Nonliving Things

In this activity, you will watch a series of videos that show different changes that occur in living and nonliving things.

Procedures and Questions

1. Your teacher will show two videos. Watch each video and record what you see happening in Table 1.1. Be sure to include any observations that suggest change is occurring.

Table 1.1. Changes in Living Things

	Observations
Video 1– Puppy Growth	*New materials are made when the puppy grows, such as fur, skin, bones, and muscle.* *The dog gets bigger (gains weight/mass) as it grows.*
Video 2– Corn Growth	*New materials are made when the corn grows, such as leaves, stem, and fruit.* *The corn plant gets bigger (gains mass) as it grows.*

2. Discuss these questions with your class. Use your observations from Table 1.1 to support your ideas.

   a. What matter is there before each change takes place?

   b. Do you think any new matter is being made during each change?

   c. If so, what is the new matter and where do you think it comes from?

Teacher Talk and Actions	
Activity 1	**(25 min)**

Whole class:

Complete Step 1 together.

For Video 1—Puppy growth:

- Show Video—*Puppy Grows Into a Dog*.
- Students record their observations in Table 1.1.

For Video 2—Corn growth:

- Show Video—*Corn Growth*.
- Students record their observations in Table 1.1.

Lead discussion. Invite students to share ideas about the discussion questions (Step 2).

The goal is primarily eliciting and probing student ideas here. Ask questions like these:

- Why do you think so?
- How do you think that happens?
- What evidence do you have for that idea from the video?

Do not make the connections for students at this point if they struggle to see the similarities between changes in living things.

3. Your teacher will show two more videos. Watch each video and record what you see happening in Table 1.2. Be sure to include any observations that suggest change is occurring.

Table 1.2. Changes in Nonliving and Living Things

	Observations
Video 1– Nylon Formation	*A new material (a white, threadlike material) comes out of the spider's body.* *As more of the white, threadlike material is produced, the web grows (gets heavier/more massive).*
Video 2– Spider Spinning Silk	*New material (a white, stringy material) is made at the interface of the two liquids.* *The white, stringy material probably gains weight/mass as it grows.*

4. Discuss these questions with your class. Use your observations from Table 1.2 to support your ideas.
   a. What matter is there before each change takes place?
   b. Do you think any new matter is being made during each change?
   c. If so, what is the new matter and where do you think it comes from?
   d. What do you think has happened to the matter that was there in the beginning?

Teacher Talk and Actions
Activity 1 (cont)
Whole class: Complete Step 3 together. For Video 1—Nylon formation: • Show Video—*Hexamethylenediamine and Adipic Acid.* • Students record their observations in Table 1.2. For Video 1—*Spider Spining Silk*: • Show Video—*Spider Spinning Silk.* • Students record their observations in Table 1.2. Lead discussion. Invite students to share ideas about the discussion questions (Step 4). The goal is primarily eliciting and probing student ideas here. Ask follow-up questions like these: • Why do you think so? • How do you think that happens? • What evidence do you have for that idea from the video? Do not make the connections for students at this point if they struggle to see the similarities between changes in nonliving and living things. They will have a chance to revisit their ideas later in the unit.

Pulling It Together

Work on your own to answer the question. Be prepared for a class discussion.

1. After observing and discussing the changes shown in the videos, how would you answer the Key Question, **How are changes in the matter that makes up living and nonliving things similar?**

   In both cases, new materials may be made or "grow." The new materials might have come from other materials that are used up. When new materials are made or "grow," there may be a change (e.g., increase) in weight/mass of the "growing" thing (e.g., dog; corn plant; stringy, white material).

Teacher Talk and Actions

Pulling It Together (5 min)

Individuals:

Direct students to answer the Key Question on their own, in detail. Not all students will write something similar to the ideal response provided. At this point, they should know that the mass increases, but it is OK if they aren't sure where it comes from.

Closure and Link (2 min)

Whole class:

Summarize the lesson:

We have observed some similarities in changes that occur in living and nonliving things. One similarity we noticed is that new stuff can be made.

Link to the next lesson:

Tell students:

You will have more opportunities during the unit to think about how these changes are similar.

You already noticed that living things are made up of a lot of stuff (e.g., fur, skin, bones, muscles, and much more!). And so are nonliving things. In the rest of Chapter 1 we'll focus on simpler stuff that makes up nonliving things and the changes that stuff undergoes. The ideas we learn will eventually help us explain what happens when living things grow and repair their bodies.

Lesson Guide
Chapter 1, Lesson 1.2
Detecting New Substances

Focus: Students explore changes in matter by examining properties of starting and ending substances: first using their physical properties (appearance, conductivity, solubility) and behaviors (interactions with limewater or burning splint for gases), and then by examining cards containing information on additional properties of the substances. The lesson challenges the common idea that two substances with only one or two properties in common are the same substance.

Key Question: How do we know if a different substance has been made?

Target Idea(s) Addressed

- Each substance has a unique set of characteristic properties, such as color, odor, state of matter at room temperature, conductivity and solubility, and how it behaves during certain tests (like limewater and glowing splint tests).
- The characteristic properties of substances can be observed or measured and used to decide if two substances are the same or different.

Science Practice(s) Addressed

- Analyze and interpret data to provide evidence for phenomena.
- Apply scientific ideas and evidence to construct and analyze explanations for phenomena.

Materials

Activity 1: Video—*Hexamethylenediamine and Adipic Acid, Card Pack 1, Card Pack 2, Change Chart,* nylon samples, distilled water samples, conductivity meters, dropper bottles of distilled water, vials

Activity 2: Dry steel wool, steel wool soaking in vinegar, paper towels, flask, fingers of nitrile gloves, Card Pack 1, Card Pack 2, Change Chart, rust samples, Conductivity meters, Dropper bottles of distilled water, Vials, Video—*Testing for Oxygen Gas*, Video—*Iron Rusting*

Activity 3: Vinegar, Baking soda, Teaspoons, Measuring cup, Quart-size zipper plastic bag, *Card Pack 1, Card Pack 2, Change Chart,* Sodium acetate samples, distilled water samples, conductivity meters, dropper bottles of distilled water, vials, Video—*Testing for Carbon Dioxide Gas: Limewater Test*

Advance Preparation:

Organize materials in advance so that each set of partners or team has materials when class begins.

Phenomena, Data, or Models	Intended Observations	Purpose	Rationale or Notes
Activity 1 Observations of physical properties of starting and ending substances when hexamethylenediamine and adipic acid are mixed Cards containing characteristic property data for starting and ending substances	Nylon thread has different physical properties from hexamethylenediamine and adipic acid. These properties include things we can observe (state of matter at room temperature, color, smell, conductivity, solubility). The properties of nylon can be used to identify it and distinguish it from starting substances, hexamethylenediamine and adipic acid.	Students observe properties of some specific substances firsthand before and after they are mixed and examine substance cards for data about properties that can't be observed first hand in class.	The nylon reaction provides a simple context for polymer formation that will prepare students for later lessons that involve polymer formation in complex biological systems. Introduces students to various properties of substances, some that they can observe and others that they can find on data cards.
Activity 2 Observations of physical properties of starting and ending substances when steel wool is exposed to air Cards containing characteristic property data for starting and ending substances Video showing behavior of oxygen gas in burning splint test	Rust has different physical properties from iron and oxygen. These properties include things we can observe (state of matter at room temperature, color, smell, conductivity, solubility) and others that involve observing how the substances behave during certain tests (burning splint test for oxygen).	Students observe properties of some specific substances firsthand before and after they are mixed and examine substance cards for data about properties that can't be observed first hand in class.	The iron rusting reaction provides a simple context for a chemical reaction with a gaseous reactant. Introduces students to various properties of substances, some that they can observe and others that they can find on data cards.

Phenomena, Data, or Models	Intended Observations	Purpose	Rationale or Notes
Activity 3 Observations of physical properties of starting and ending substances when baking soda and vinegar are mixed Cards containing characteristic property data for starting and ending substances Video showing behavior of carbon dioxide gas in limewater test	Carbon dioxide, sodium acetate, and water have different physical properties from baking soda and vinegar. These properties include things we can observe (state of matter at room temperature, color, smell, conductivity, solubility) and others that involve observing how the substances behave during certain tests (limewater test for carbon dioxide).	Students observe properties of some specific substances firsthand before and after they are mixed and examine substance cards for data about properties that can't be observed firsthand in class. Students generalize across observations across the three changes before reading the generalization (Science Idea #1)	The baking soda + vinegar reaction provides a simple context for a chemical reaction with a gaseous product. Introduces students to various properties of substances, some that they can observe and others that they can find on data cards.
Activity 4 Observations of an explanation that correctly uses claim, science ideas, and evidence	Eva's claim responds to the question. Eva's explanation: • Lists the idea that different substances have different sets of properties that can be used to distinguish them (Science Idea #1) • Presents specific properties as evidence a new substance formed • Uses Science Idea #1 to link evidence to claim	Introduces students to a high-quality explanation in a simple context (substance level) and criteria for judging its quality	Introducing the explanation scaffold at the substance level will prepare them for more sophisticated explanations at the atomic/molecular level in Lesson 1.5 and beyond.

Phenomena, Data, or Models	Intended Observations	Purpose	Rationale or Notes
Pulling It Together Table containing conductivity and solubility data for starting and ending substances of another change	The ending substance (white crystals) has different properties (different conductivity and water solubility) from the starting substances (metallic gray solid, yellow gas).	Students decide that the ending substance is different from the starting substances based on its having different characteristic properties, and students justify their decision based on evidence that properties of ending substance are different from those of starting substance and the idea that if different properties then different substance (Science Idea #1).	Giving students a chance to apply the explanation scaffold in a simple context (substance level, only one science idea to draw from) will prepare them for writing more complex explanations that involve having to choose from more science ideas (Lesson 1.3 and beyond), make more inferences from data (Lesson 1.3 and beyond), and also require reasoning from models (Lesson 1.5 and beyond).

Lesson 1.2—Detecting New Substances

What do we know and what are we trying to find out?

All of the changes that you observed in Lesson 1.1—in both the living and nonliving examples—involved changes in matter. Understanding what matter is made up of is key to understanding how matter changes.

Some matter, like solid table sugar, liquid water, and oxygen gas, consists of one substance. More commonly, matter is made up of several different substances—air, for example, is a mixture mainly of nitrogen gas, oxygen gas, water vapor, and carbon dioxide gas. Likewise, the matter that composes the bodies of living organisms—dogs, trees, humans, and lizards, for example—is made up of many different substances that are organized into complex structures.

Many substances look very similar. Take water and vinegar, for example (Figure 1.2).

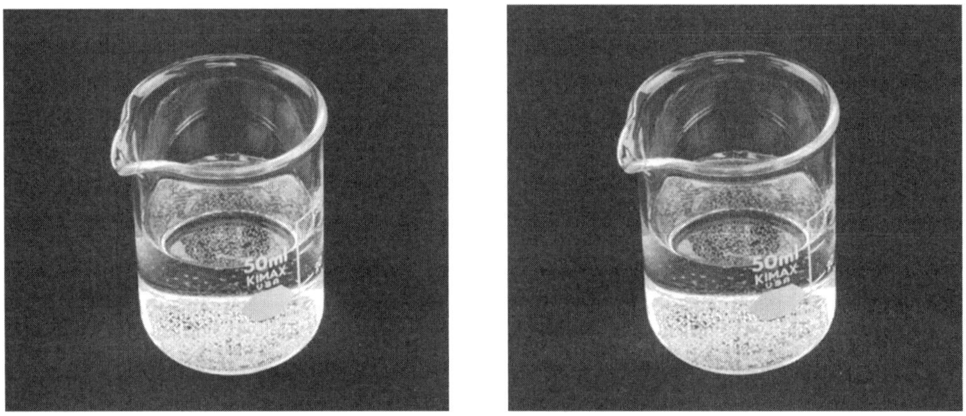

Water Vinegar

Figure 1.2. Comparing Water and Vinegar

Both are clear colorless liquids. So if you had a cup of each, how would you know which is which? If you have had experiences with vinegar, you may have noticed that it has a very strong odor, and you probably know that water doesn't have much of an odor at all. The odor of a substance is one way to tell if two substances are the same or different.

Properties like odor are called "characteristic" properties because they help define that substance. Other characteristic properties include the melting point and boiling point of a substance, whether or not a substance can be dissolved in water, and the electrical conductivity of a substance. No matter how small the sample of a substance is, what shape it is, when it is observed, or where it is located, a substance's characteristic properties will always be the same. And even though different substances may have **some** characteristic properties that are the same, no two substances have characteristic properties that are **all** the same. For this reason, scientists can use the set of characteristic properties of a substance to identify it when they don't know what it is.

Teacher Talk and Actions
What do we know and what are we trying to find out? (3 min)
Whole class: Read text under "What do we know and what are we trying to find out?" aloud to the class.

In this lesson, you will explore the characteristic properties of the substances that were involved in some of the changes you saw in the videos in Lesson 1.1. This will help you make sense of what is happening during each of these changes.

Answer the Key Question to the best of your knowledge. Be prepared to share your ideas with the class.

Key Question: How do we know if a different substance has been made?

We might see something that is a different color or we might see bubbles. In the previous lesson's video we saw the scientist pour two liquids together and pull out a solid thread.

Teacher Talk and Actions	
Key Question	**(5 min)**

Individuals:

Students respond to the Key Question: **How do we know if a different substance has been made?**

Note that the ideal student response is only one example of how students might be expected to respond at this point in the unit.

Whole class:

Lead discussion. Invite students to share ideas about the Key Question.

The goal is primarily eliciting and probing student ideas.

When students share their answers, do not indicate if their answers are right or wrong. However, do not let students assume by your silence that they were correct. Make it clear that students are expressing ideas that may change based on what they will see in the lessons. (Remind students of this regularly.)

Guide students in thinking about instances in other classes, such as Earth science, where they may have "tests" to identify substances. If something always behaves a certain way in a test, that might help us distinguish substances.

Activity 1: Investigating a Change: Hexamethylenediamine and Adipic Acid

Materials
For each team of students
Card Pack 1 (green outline)
Card Pack 2 (yellow outline)
Change Chart

In this activity, you will investigate the first of three different examples of changes in matter. You will refer to these three changes often during the unit, so it is important that you make very careful observations.

To help in your investigations, you will look at charts that represent the three changes and cards that describe the starting and ending substances that are involved. Each card provides data on the characteristic properties for a substance. You will use these cards and charts to think about how we can tell if a different substance has been made.

Procedures and Questions

1. As a whole class, brainstorm a list of words to describe what you are able to observe about substances—their physical state, appearance, or smell, for example. Your teacher will record this list on the board or chart paper so that you can use it whenever you write descriptions of substances. If you think of more ways to describe substances during the activity, add them to the list. You can use the space below to brainstorm.

Teacher Talk and Actions	
Activity 1	**(40 min)**

Whole class:

Complete Step 1 as a whole class. Chart descriptors in a permanent location so students can refer to the list in future lessons. Help students think about how they would describe everyday things: state, color, luster (shiny/dull), smell (none/pungent/fishy), transparent/opaque, and so forth.

2. Notice the labels at the top of each column in Table 1.3 (p. 15). In this activity, you will be writing information in the appropriate boxes in some or all of the columns. To get started, write the names for the starting substances involved in the change you are investigating. Hint: Look at the title of the table.

3. Observe the video of the starting substances and fill in whatever information you now have about the substances in the appropriate columns in rows 1 and 2 of Table 1.3.

4. Using *Card Pack 1* and the *Change Chart,* do the following:

    a. Find the cards for Hexamethylenediamine and Adipic Acid. Place them on the Starting Substances (left) side of the *Hexamethylenediamine and Adipic Acid Change Chart.*

    b. Compare the observations you recorded for starting substance 1 and starting substance 2 in Table 1.3 with the properties on the Hexamethlyenediamine and Adipic Acid cards. Fill in the table with information from the data cards. Does the information on the cards match your observations? What else can you learn about these substances from the cards?

Starting Substance 1 (Hexamethylenediamine): We observed a clear, colorless liquid. The card indicates that hexamethylenediamine is a white, powdery solid that dissolves in water to form a clear, colorless liquid. So what we observed on the video was probably the solid dissolved in water. The card also indicates it has a fishy smell, which we couldn't detect on the video. And the card provides data about its melting point (42 °C), density (.93 g/cm^3), and its conductivity.

Starting Substance 2 (Adipic Acid): We observed a clear, colorless liquid. The card indicates that adipic acid is a white, powdery solid that dissolves in water to form a clear, colorless liquid. So what we observed in the video was probably the solid dissolved in water. The card also indicates that it is odorless, has a melting point of 152 °C, and density of 1.36 g/cm^3. So several of its properties are different from the properties of hexamethylenediamine.

5. Watch as your teacher presents a video of the change. Make careful observations and write detailed descriptions of what you observed during the change in row 3 of Table 1.3. Use words from the list you brainstormed in Step 1.

6. The starting substances in this investigation are too dangerous to have in the classroom. However, the ending substances are safe. Ask your teacher for pure samples of the ending substances.

7. Observe each ending substance. Record observations about the ending substance's state of matter (at room temperature), color, and smell in Table 1.3.

    When smelling a substance, remove the cap and wave your hand over the opened container to waft the odor toward your nose. Never smell a substance directly.

Teacher Talk and Actions
Activity 1 (cont)
Whole class: Show video of *Hexamethylenediamine and Adipic Acid*. Pause the video before the scientist starts spooling a white, stringy substance (which students will soon learn is nylon). Remember to turn the sound off. After examining the data cards, students will learn that substances 1 and 2 are solids at room temperature and that they dissolve in water. Ask students how they can reconcile observing a liquid when the card says it is a solid. Once students understand that they were really observing a solution, it's OK for them to cross out liquid and write in solid for both starting substances 1 and 2. After students have completed Step 4b, turn the video on again (without sound). Step 5: Students may notice that the white, stringy substance is moist and conclude that water is an ending substance. If so, ask,"What evidence would we need to support the conclusion that water is an ending substance?" Hand out samples of ending substance 1 (nylon) and ending substance 2 (water), along with the conductivity meters, dropper bottles of water, and extra vials (for testing solubility). Note: The water must be distilled or deionized. Tap water has trace minerals and/or salts in it, which will cause a false positive when testing with the conductivity meter. Distilled water is available at grocery stores. Demonstrate the wafting technique for smelling substances. Encourage students to record smell only if it is very obvious.

8. Test each ending substance's conductivity by touching both leads (metal prongs) of the conductivity meter to the substance. Record whether each substance is conductive in rows 4 and 5 of Table 1.3 in the Conductive column.

   If the bulb or diode lights or flashes, electricity is being conducted through the sample and the substance is considered conductive.

9. Place a small amount of the ending substance into a vial. Test its solubility by filling the vial halfway with water. Record whether each substance is soluble in rows 4 and 5 of Table 1.3 in the Soluble column.

10. If you observed that the substance is soluble, test its conductivity again. Record your observation for each substance in rows 4 and 5 of Table 1.3 in the Conductive When Dissolved column. If the substance is not soluble, write N/A.

11. Using *Card Pack 2,* do the following:

    a. Compare your observations of the ending substances that you recorded in Table 1.3 with the properties of the different substances shown in *Card Pack 2* for the Hexamethylenediamine and Adipic Acid investigation.

    b. Now look at the *Hexamethylenediamine and Adipic Acid Change Chart*. Match any substance card(s) that you can to this change. Place them on the Ending Substances (right) side of the chart. Write the name(s) of the ending substance(s) in the first column of rows 4 and 5 of Table 1.3.

Teacher Talk and Actions
Activity 1 (cont)
Demonstrate how to use the conductivity meter. Demonstrate how to test for solubility. Since nylon does not form a clear solution, it is not soluble. If a substance is soluble, it will dissolve to form a clear solution mixed with water. If the solution is cloudy the substance is slightly soluble. Steps 7–10: Students should observe that white, stringy solid is not conductive and is not soluble in water and, from this data, conclude that ending substance 1 is probably nylon. Students should observe that the clear, colorless liquid (ending substance 2) is odorless and not conductive. From this data, they can conclude that ending substance 2 is water. They can also conclude that the clear, colorless liquid they observed on the nylon thread being pulled from the beaker is probably water. To be sure, they would have to test its conductivity. Since both hexamethylenediamine and adipic acid are conductive when dissolved in water, the finding that the clear, colorless liquid on the nylon is not conductive would provide evidence that it is NOT either of the starting substances. Step 11: Students have evidence to place the four cards on the change mat. At this point, it's OK if they're not sure about whether or not water is an ending substance. After learning about atomic composition of the starting and ending substances in Lesson 1.5, they will be able to reason that the H from hexamethylenediamine and OH from adipic acid linked together to form water as the monomers were linked to form nylon.

Activity 2: Investigating a Change: Steel Wool (Iron) and Air (Oxygen)

Materials
For each team of students
Dry steel wool
Steel wool soaking in vinegar
Paper towels
Flask
Finger of a nitrile glove
Card Pack 1 (green outline)
Card Pack 2 (yellow outline)
Change Chart

In this activity, you will investigate another change using your own observations and information on the substance cards and *Change Chart*.

Procedures and Questions

1. Write the names for the starting substances involved in this change in Table 1.4 (p. 16).

2. Observe the dry steel wool. Record your observations about the steel wool's state of matter, color, and smell in the appropriate columns of Table 1.4.

3. Test the steel wool's conductivity by touching both leads of the conductivity meter to the substance. Record whether the substance is conductive in Table 1.4.

4. Place a small amount of steel wool into a vial. Test its solubility by filling the vial halfway with water. Record whether the substance is soluble in Table 1.4.

5. If you observed that the substance is soluble, test its conductivity again. Record your observation in Table 1.4. If the substance is not soluble, write N/A.

6. Steel wool has a protective coating on it that must be removed to expose the iron contained in the thin strands of steel wool. Soaking the steel wool in vinegar removes the protective coating. Vinegar is not considered a starting substance in this case, but it helps the change occur. Remove a small piece of steel wool from the vinegar. Squeeze out most of the vinegar and blot off the rest with paper towels.

7. Pull apart the strands so that more steel wool is exposed to the air and then place the steel wool in a clean, dry flask. Immediately cover the mouth of the flask with one finger cut from a nitrile glove. Take note of how the finger of the glove looks now.

8. Observe the air around you. Record your observations about the air's state of matter, color, and smell in the appropriate columns in Table 1.4. Because air is made up of several different gases, how can we find out if there is oxygen in the air?

Teacher Talk and Actions	
Activity 2	**(40 min)**
(Intentionally left blank)	

9. Watch as your teacher shows a demonstration of a test for the presence of oxygen.

   When oxygen gas is present, a burning splint will continue to burn.

10. Using *Card Pack 1* and the *Change Chart,* do the following:

    a. Find the cards for Steel Wool (Iron) and Air (Oxygen). Place them on the Starting Substances (left) side of the *Steel Wool (Iron) and Air (Oxygen) Change Chart.*

    b. Compare the observations that you recorded for steel wool and air in Table 1.4 with the properties on the Iron and Oxygen cards. Fill in the table with information from the data cards. Does the information on the cards match your observations? What else can you learn about these substances from the cards?

    Steel wool (Iron): We observed its appearance (solid at room temperature, metallic gray color), its conductivity (yes), and its water solubility (no). The data card provided information about its melting point and density.

    Air (Oxygen): We observed its appearance (gas at room temperature, colorless), and its behavior in splint test (positive). The data card provided information about its melting point and density.

11. Watch as your teacher presents a video of the change you are investigating. Make careful observations and write detailed descriptions of what you observe during the change in Table 1.4. Use words from the list you brainstormed in Step 1 of Activity 1.

12. Ask your teacher for a pure sample of the ending substance. Also note that the ending substance is being formed in your flask.

13. Observe the ending substance. Record observations about the ending substance's state of matter (at room temperature), color, and smell in Table 1.4.

14. Test the ending substance's conductivity by touching both leads (metal prongs) of the conductivity meter to the substance. Record whether the substance is conductive in Table 1.4 in the Conductive column.

15. Place a small amount of the ending substance into a vial. Test its solubility by filling the vial halfway with water. Record whether the substance is soluble in Table 1.4 in the Soluble column.

16. If you observed the ending substance to be soluble, test its conductivity again. Record whether the substance is conductive when dissolved in Table 1.4 in the Conductive When Dissolved column. If the substance is not soluble, write N/A.

17. Using *Card Pack 2,* do the following:

    a. Compare your observations of the ending substance to the properties of the different substances in *Card Pack 2* for the Steel Wool (Iron) and Air (Oxygen) investigation.

    b. Match any substance card(s) that you can to this change. Place them on the Ending Substances (right) side of the *Steel Wool (Iron) and Air (Oxygen) Change Chart.* Write down the name of the ending substance in Table 1.4.

Teacher Talk and Actions

Activity 2 (cont)

Show Video—*Testing for Oxygen Gas*

A glowing splint will usually relight only when in an enriched-oxygen environment (e.g., oxygen gas produced from electrolysis). For this reason, a burning splint continuing to burn/going out is a better indicator of oxygen gas presence/absence during the rusting reaction.

Show Video—*Iron Rusting*

Give out samples of rust once students have finished watching the video.

Activity 3: Investigating a Change: Baking Soda and Vinegar

Materials
For each team of students
Vinegar
Baking soda
Teaspoon
Measuring cup
Quart-size zipper plastic bag
Card Pack 1 (green outline)
Card Pack 2 (yellow outline)
Change Charts

In this third change, it is important to keep track of all of the substances that are involved.

Procedures and Questions

1. Write the names for the starting substances involved in this change in Table 1.5 (p. 17).
2. Observe the baking soda. Record your observations about baking soda's state of matter, color, and smell in the appropriate columns in Table 1.5.
3. Test baking soda's conductivity by touching both leads of the conductivity meter to the substance. Record whether the substance is conductive in Table 1.5.
4. Place a small amount of baking soda into a vial. Test its solubility by filling the vial halfway with water. Record whether the substance is soluble in Table 1.5.
5. If you observed that the substance is soluble, test its conductivity again. Record your observation in Table 1.5 in the Conductive When Dissolved column. If the substance is not soluble, write N/A.
6. Repeat Steps 2–5 for vinegar.
7. Using *Card Pack 1* and the *Change Chart,* do the following:
   a. Find the cards for Baking Soda and Vinegar. Place them on the Starting Substances (left) side of the *Baking Soda and Vinegar Change Chart.*
   b. Compare the observations that you recorded in Table 1.5 to the properties on the Baking Soda and Vinegar cards. Fill in the table with information from the data cards. Does the information on the cards match your observations? What else can you learn about these substances from the cards?

   We observed that baking soda is a white solid at room temperature that dissolves in water and conducts when it is dissolved. The data card provided us with information about its melting point (60 °C) and density (2.16 g/cm^3).

   We observed that vinegar is a clear, colorless, smelly liquid at room temperature that conducts electricity and is soluble in water. The data card provided us with information about its melting point (17 °C) and density (1.05 g/cm^3).

Teacher Talk and Actions	
Activity 3	**(40 min)**
(Intentionally left blank)	

8. Put a teaspoon of baking soda directly into the quart-size bag. Measure 25 ml of vinegar in a measuring cup.

9. Place the cup containing vinegar upright in the plastic bag. Take care that you do not spill the contents of the cup. Push as much air out of the plastic bag as possible and seal the bag. Observe how the bag looks and feels now, and record your observations in Table 1.5. Then, without opening the bag, tip over the cup of vinegar in the bag so that the vinegar mixes with the baking soda. Make careful observations of what happens in the bag during the change.

10. Write a detailed description of what happened inside the bag during the change in row 3 of Table 1.5. Write about any changes in how the bag looks and feels now compared with how it looked and felt when you completed Step 9.

11. Watch as your teacher shows a demonstration of the limewater test.

    If a colorless limewater solution turns milky white (test result is positive), it means that the limewater solution has interacted with carbon dioxide gas. In other words, the sample has tested positive for carbon dioxide gas.

12. Record your observations about ending substance 1 in Table 1.5.

13. Ask your teacher for pure samples of the ending substances 2 and 3.

14. Observe ending substances 2 and 3. For each ending substance, record observations about its state of matter (at room temperature), color, and smell in Table 1.5.

15. Test the conductivity of ending substances 2 and 3 by touching both leads (metal prongs) of the conductivity meter to the substance. Record whether each substance is conductive in Table 1.5 in the Conductive column.

16. Place a small amount of each ending substance into two separate vials. Test the solubility of each by filling the vial halfway with water. Record whether each substance is soluble in Table 1.5 in the Soluble column.

17. If you observed either of the ending substances to be soluble, test its conductivity again. Record whether the substance is conductive when dissolved in Table 1.5 in the Conductive When Dissolved column. If the substance is not soluble, write N/A.

18. Using *Card Pack 2* and the *Change Chart,* do the following:

    a. Compare your observations of the ending substances with the properties of the different substances in *Card Pack 2* for the Baking Soda and Vinegar investigation.

    b. Match any substance card(s) that you can to this change. Place them on the Ending Substances (right) side of the *Baking Soda and Vinegar Change Chart*. Write down the name of the ending substance(s) in Table 1.5.

Teacher Talk and Actions
Activity 3 (cont)
Show Video—*Testing for Carbon Dioxide Gas: Limewater Test* Science Notes to Teacher: The white solid (suspended in liquid) that forms during a positive limewater test is calcium carbonate (chalk). Note that you will not be giving them samples of carbon dioxide. Explain that carbon dioxide is another substance in the air. They can write down observations such as state of matter, color, and odor. Give the students samples of sodium acetate and water once they have completed Step 12.

Table 1.3. Hexamethylenediamine and Adipic Acid Investigation

Starting Substances						
Name	State of Matter (at room temperature)	Color	Smell	Conductive	Soluble	Conductive When Dissolved
1. Starting Substance 1 *Hexamethylene-diamine*	*~~liquid~~ solid*	*clear*	*Fishy smell (from data card)*		*Yes (from data card)*	*Yes (from data card)*
2. Starting Substance 2 *Adipic Acid*	*~~liquid~~ solid*	*clear*	*Odorless (from data card)*		*Yes (from data card)*	*Yes (from data card)*
What did you observe during the change?						
3. *At first, there were two clear, colorless liquids. When they were mixed, a white thread-like substance was pulled from where the two liquids were in contact (interface).*						
Ending Substances						
Name	State of Matter (at room temperature)	Color	Smell	Conductive	Soluble	Conductive When Dissolved
4. Ending Substance 1 *Nylon*	*solid*	*white*	*none, odorless*	*no*	*no*	*N/A*
5. Ending Substance 2 *Water*	*liquid*	*clear*	*none, odorless*	*no*	*yes*	*no*

Table 1.4. Steel Wool (Iron) and Air (Oxygen) Investigation

Starting Substances						
Name	State of Matter (at room temperature)	Color	Smell	Conductive	Soluble	Conductive When Dissolved
1. Starting Substance 1 *Steel Wool (Iron)*	*Solid*	*Gray*	*None*	*Yes*	*No*	*N/A*
2. Starting Substance 2 *Air (Oxygen)*	*Gas*	*Colorless*	*None*			
What did you observe during the change?						
3. *A reddish orange solid formed and the balloon was pushed into the flask.*						
Ending Substance						
Name	State of Matter (at room temperature)	Color	Smell	Conductive	Soluble	Conductive When Dissolved
4. Ending Substance 1 *Rust (Iron Oxide)*	*Solid*	*Reddish orange*	*None*	*No*	*No*	*N/A*

SE p. 16

Table 1.5. Baking Soda and Vinegar Investigation

Starting Substances						
Name	State of Matter (at room temperature)	Color	Smell	Conductive	Soluble	Conductive When Dissolved
1. Starting Substance 1 *Baking Soda*	*Solid*	*White*	*None*	*No*	*Yes*	*Yes*
2. Starting Substance 2 *Vinegar*	*Liquid*	*Clear*	*Fishy, pungent*	*Yes*	*Yes*	*Yes*
What did you observe during the change?						
3. *Bubbles formed and the bag puffed up. At the end, there was just a clear, colorless liquid in the bag.*						
Ending Substances						
Name	State of Matter (at room temperature)	Color	Smell	Conductive	Soluble	Conductive When Dissolved
4. Ending Substance 1 *Carbon Dioxide*	*Gas*	*Colorless*				
5. Ending Substance 2 *Sodium Acetate*	*Solid*	*White*	*None*	*No*	*Yes*	*Yes*
6. Ending Substance 3 *Water*	*Liquid*	*Colorless*	*None*	*No*	*Yes*	*No*

Teacher Talk and Actions

Activity 3 (cont)

Information for the next teacher page (p. 45; table to be projected/filled out):

Discuss: Do any two of the substances have all the same characteristic properties? What does this tell you about the two substances?

Give reasons for your answer. A set of characteristic properties gives a substance its identity. So if two substances don't have the same set of characteristic properties, they can't have the same identity. For example, I have a certain birthdate, place of birth, eye color, blood type, etc. While other people might have some of the same characteristics, no one has the same set (or they would be me).

What does this tell you about the two substances?

Help students reach consensus that no two substances are the same because they do not have all the same properties.

19. With your class, fill out the table that your teacher is going to project. Then answer the following questions:

    a. Compare the properties of the substances. Do ANY of the substances have exactly the same set of properties? Give an example from data on the substance cards or from your observations to support your answer.

None of the substances have the same properties. For example, steel wool is conductive, but rust isn't.

    b. Did substances that had a similar appearance have the same set of properties? Give an example from data on the substance cards or from your observations to support your answer.

None of the substances have the same properties, even the ones that were similar. For example, water and vinegar are both clear, colorless liquids, but vinegar has an odor, while water does not. So while they are similar, they don't have exactly the same set of properties.

Name	State of Matter (at room temperature)	Color	Smell	Conductive	Soluble	Boiling Point
Hexamethylene-diamine						
Adipic Acid						
Nylon						
Water						
Steel Wool (Iron)						
Air (Oxygen)						
Rust (Iron Oxide)						
Baking Soda						
Vinegar						
Carbon Dioxide						
Sodium Acetate						

Science Ideas

Science ideas are accepted principles or generalizations about how the world works based on a wide range of observations and data collected and confirmed by scientists. Because these science ideas are consistent with the available evidence, you are justified in applying science ideas to other relevant observations and data.

Activities 1, 2, and 3 were intended to help you understand an important idea about substances and their properties. Read the idea below. Look back through Lesson 1.2. In the space provided after the science idea, give evidence that supports the idea.

Science Idea #1: Every substance has a unique set of characteristic properties, such as color, odor, density, melting point, conductivity, solubility, and how it behaves (such as in the limewater and burning splint tests). The properties of substances can be observed or measured and used to decide if two substances are the same or different.

Evidence (from your observations or data on the substance cards):

Water and vinegar are both clear, colorless liquids. However, water and vinegar have different odors, conductivities, melting points, and densities. Because they don't have the same set of properties, water and vinegar are different substances.

Teacher Talk and Actions
Science Ideas **(10 min)**
Tell students: Compare your new ideas about substances with scientists' ideas. Read the science idea aloud to the class. Chart the science idea so that students can refer to it during the unit. Use large print so it can be seen anywhere in the class. Give students time to write down at least one piece of evidence that supports the idea.

Activity 4: Constructing Scientific Explanations—Part I

Scientists try to explain how the world works using evidence from their observations, science ideas, and logical reasoning that links the evidence and the science ideas. Throughout this unit, you too will be asked to make observations and explain them using evidence, science ideas, and logical reasoning. This activity introduces you to what a scientific explanation needs to include and how to evaluate its quality.

Procedures and Questions

1. After completing Activities 1, 2, and 3, students were asked whether a new substance was formed when hexamethylenediamine and adipic acid were mixed and to explain their claims. Eva looked at the cards for hexamethylenediamene, adipic acid, and nylon and decided that a new substance was formed.

   To help write a valid explanation of her claim, Eva organized her thinking as shown in Table 1.6. Examine the different parts of her explanation and read the description of each part.

Table 1.6. Parts of an Explanation

	What It Is	Eva's Explanation
Question	The question to be answered	
Claim	A statement or conclusion that is intended to respond to the question	*Yes, a new substance formed when hexamethylenediamine and adipic acid were mixed.*
Science Ideas	Widely accepted scientific ideas, concepts, or principles that can be used to show why the evidence supports the claim	*Every substance has its own set of properties (Science Idea #1).*
Evidence	Data that are relevant to and support the claim and can be confirmed by others Data are based on observations about the world that are either made with our senses or measured with instruments.	*Hexamethylenediamine and adipic acid are clear, colorless liquids and soluble in water.* *Nylon is a white, fibrous solid and not soluble in water.*

Teacher Talk and Actions	
Activity 4	**(30 min)**
Whole class: Read the introductory paragraph aloud to the class. Individuals: Students independently review the parts of an explanation and the sample.	

Eva's explanation: *Yes, a new substance formed when hexamethylenediamine and adipic acid were mixed. Every substance has its own set of properties. If two substances have different properties, they are different substances. Properties can be used to distinguish one substance from another (Science Idea #1). The most likely ending substance from the card pack, nylon, has a different set of properties than the two starting substances, hexamethylenediamine and adipic acid. For example, both hexamethylenediamine and adipic acid are clear, colorless liquids that are both soluble in water, whereas nylon is a white, fibrous solid that is not soluble in water. Because nylon has a different set of properties than the two starting substances, it is a different substance.*

Here are some questions that can help us decide whether Eva's explanation is valid. Examine Eva's explanation and answer the questions about each part. Discuss whether or not Eva's explanation answers each question, and record any ideas that come up during the discussion.

Explanation Quality Questions

- Does the **claim** respond to the question?
- Are the **science ideas** that are listed relevant to the claim?
- Does the **evidence** support the claim?
- Are the **science ideas** used to show why the evidence supports the claim?

If the answer to any of these questions is no, then the explanation is not valid. You can use these criteria to evaluate and improve the quality of your own explanations and also to evaluate other people's explanations.

In later lessons, you will learn about other parts that are needed for a valid explanation and additional questions you can use to evaluate the quality of explanations.

Teacher Talk and Actions

Activity 4 (cont)

Individuals:

Students read Eva's explanation.

Whole class:

Lead discussion about the Explanation Quality Questions. Invite students to share answers to the Explanation Quality Questions for Eva's explanation and ask them to give reasons for their judgments. Look to see if students are giving reasons why Eva's explanation meets the quality questions, such as the following:

- The question asks if a new substance formed and Eva's claim is about whether a new substance formed.
- The question asks if the science ideas are relevant to the claim and Eva's explanation lists a science idea that can be used to decide if a new substance formed.
- Eva cites evidence about the properties of starting substances and ending substances and summarizes the evidence by stating that the properties of the starting and ending substances are different, which she can then use to support her claim.
- Eva's explanation uses the word *because* to link the science idea to the evidence.

Pulling It Together

Work on your own to answer these questions. Be prepared for a class discussion.

1. Now that you have learned something about the starting substances and ending substances for the three changes, how would you answer the Key Question, **How do we know if a different substance has been made?**

Every substance has its own unique set of characteristic properties, so if two substances have different properties they are different substances. To determine if a different substance has been made, we need to determine its characteristic properties. Some properties, like appearance and state of matter at room temperature, can be observed. Other properties, like melting point and density, can be measured. If the set of properties of the substance is different from the set of properties of the starting substances, then a different substance has been made.

2. When grayish metal was added to a container of yellow gas, white crystals appeared inside the container. Properties of the starting substances and of the white crystals that form were observed and measured by a scientist and are shown in Table 1.7.

Table 1.7. Gray Metal and Yellow Gas

Appearance	Conductivity	Soluble in Water	Conductive When Dissolved
Metallic, grayish solid	High	No	N/A
Yellow gas	Not conductive	Yes	Yes
White crystals	Low	Yes	Yes

Teacher Talk and Actions
Pulling It Together **(10 min)**
Individuals: Direct students to answer the questions on their own, in detail. Whole class: Lead discussion. Invite students to share ideas about the Key Question. Direct students to look back at their initial ideas about the Key Question and consider how their ideas have changed. Pulling It Together Question 2 can be used as an embedded assessment. • Poll students to see who thinks any two of the substances are the same, and who thinks none of the substances are the same. • Ask for volunteers to explain their thinking. You will want to review students' explanations after class. During class, you may want to ask students: • Is the science idea you listed relevant to your answer? • Is all the evidence consistent with the science idea? • How did you use the science idea in your explanation?

a. Was a different substance made when the grayish metal was mixed with the yellow gas? Use the table below to record notes before writing an explanation. Remember to use the Explanation Quality Questions as a guide.

Question	Was a different substance made when the grayish metal was added to the container of yellow gas?
Claim	*Yes. A new substance formed.*
Science Ideas	*Every substance has a unique set of characteristic properties that can be observed or measured and used to decide if two substances are the same or different (Science Idea #1).*
Evidence	*The white crystals have different conductivity and solubility from both of the starting substances.*

Explanation:

Yes. A new substance formed when the grayish metal was added to the yellow gas. Every substance has a unique set of characteristic properties that can be observed or measured and used to decide if two substances are the same or different (Science Idea #1). The white crystals have different conductivity (low) compared with the metallic, grayish solid (high) and yellow gas (not conductive). Because the white crystals have different properties from the starting substances, it is a different substance.

Teacher Talk and Actions
Closure and Link **(2 min)**
Whole class: Summarize the lesson: We learned from firsthand observations and from data based on scientists' measurements that every substance has a unique set of characteristic properties. Link to the next lesson: We will learn in the next lesson where the new substances come from.

Lesson Guide
Chapter 1, Lesson 1.3
Making New Substances From Other Substances

Focus: In this lesson, students explore three changes again, but looking for specific evidence linking the increase in amounts of products with the decrease in amounts of reactants. The lesson challenges several common ideas about the relationship (or lack thereof) between starting substances and new substances that form in chemical reactions.

Key Question: As a new substance forms, what happens to the starting substances?

Target Idea(s) Addressed

- Changes during which starting substances interact to form new substances are called chemical reactions.
- During a chemical reaction, the amounts of the reactants decrease as the amounts of the products increase.

Science Practice(s) Addressed

- Analyze and interpret data to provide evidence for phenomena.
- Apply scientific ideas and evidence to construct and evaluate an explanation for real-world phenomena.

Materials

Activity 1: Vinegar, baking soda, teaspoons, measuring cups, quart-size zipper plastic bags, Video—*Hexamethylenediamine and Adipic Acid*, Video—*Steel Wool (Iron) and Air With Splint Test*

Advance Preparation

Organize materials in advance so that each set of partners or table teams has materials when class begins.

Phenomena, Data, or Models	Intended Observations	Purpose	Rationale or Notes
Activity 1 Three changes in matter: • Nylon formation from hexamethylenediamine and adipic acid • Rust formation from iron and oxygen • Carbon dioxide, sodium acetate, and water formation from baking soda and vinegar	As more and more nylon thread is pulled from the beaker, the level of the liquids decreases. As the amount of rust increases, the amounts of iron and oxygen decrease. When carbon dioxide forms, the smell (of vinegar) decreases.	Students have the opportunity to gather evidence that the ending substances of a chemical reaction actually come from the starting substances. (Note: these are qualitative data, based on observations, not based on mass measurements of reactants and products.) Students have the opportunity to generalize across observations across the three changes before reading the generalization (Science Idea #2). (Note: More able students should note that there is a correlation between the increase in amount of product(s) and the decrease in amount of reactant(s).)	Nylon formation is the most obvious example of the amounts of starting substances decreasing as the amount of new substances forming increase; that is why it is placed first in this lesson. This activity challenges commonly held incorrect ideas that • reactants and products of chemical reactions are not related in any way and • products are just a mixture of the starting substances, not a new substance. Having students start to think about amounts is intended to prepare students for quantitative evidence (mass changes) they will encounter and use in Chapters 2–4.
Activity 2 Observations of an incomplete and partly incorrect explanation	Jeremy's claim only partly responds to the question. Evidence "the properties change" is not consistent with the idea that chemical reactions produce new substances with different properties (Science Idea #2).	Students have the opportunity to critique a flawed explanation based on its having an incomplete claim (because it doesn't state that carbon dioxide is a new substance) and on its providing evidence that isn't consistent with the science idea (substances don't change their properties).	Another explanation scaffold in a simple context is intended to prepare students for writing more complex and sophisticated explanations in Lesson 1.5 and beyond.

Lesson 1.3—Making New Substances From Other Substances

What do we know and what are we trying to find out?

In Lesson 1.2, you investigated three changes in which different substances were made by mixing other substances together. You learned that in order to determine if a new substance forms during a change, you should measure the characteristic properties of the substance.

You may not know it, but changes that produce new and different substances are occurring all around you every day. These changes are even happening in your own body right now, just like they are happening in every living thing. Can you think of any changes in matter that you have observed in your everyday life that might involve making new substances? For example, when you bake a cake, the cake has different properties from the ingredients used to make it. Where does the cake come from? What happens to the original ingredients you mixed to make the cake?

Answer the Key Question to the best of your knowledge. Be prepared to share your ideas with the class.

> **Key Question: As a new substance forms, what happens to the starting substances?**

I think the starting substances must somehow turn into the ending substances, but I don't know how that could happen.

Teacher Talk and Actions

What do we know and what are we trying to find out? (3 min)

Whole class:

Read text under "What do we know and what are we trying to find out?" aloud to the class.

Key Question (5 min)

Individuals:

Students respond to the Key Question: **As a new substance forms, what happens to the starting substances?**

Whole class:

Lead discussion. Invite students to share ideas about the Key Question.

The goal is primarily eliciting and probing student ideas here.

When students share their answers, do not indicate if their answers are right or wrong. However, do not let students assume by your silence that they were correct. Make it clear that students are expressing ideas, but those ideas may change based on what they will see in the lessons. (Remind students of this regularly.)

Challenge ideas that are inconsistent with learning from previous lessons.

Students may suggest examples that do not involve a new substance forming during a chemical reaction. Do not correct them at this point.

Activity 1: Observing Three Changes Again

Materials
For each team of students
Vinegar
Baking soda
Teaspoon
Measuring cup
Quart-size zipper plastic bag

In this activity, you will again investigate the three changes from Lesson 1.2. Your teacher will show videos of two changes. You will investigate the other change in teams at your tables. This time as you observe each change, look for changes in the amounts of starting and ending substances to see if you can figure out where the new substances come from.

Procedures and Questions

1. Watch the video of the change involving hexamethylenediamine and adipic acid. Look closely at how the amounts of starting and ending substances are changing from the beginning to the end of the video. Record the names of the substances and your observations in Table 1.8.

Table 1.8. Hexamethylenediamine and Adipic Acid

Describe what happens to the amounts of the starting substances.	
Name	What happens to the amount?
Hexamethylenediamine	It decreases.
Name	What happens to the amount?
Adipic Acid	It decreases.
Describe what happens to the amounts of the ending substances.	
Name	What happens to the amount?
Nylon	It increases.
Name	What happens to the amount?
Water	It increases.

Teacher Talk and Actions	
Activity 1	**(30 min)**

Whole class:

Read introductory text.

Remind students to focus on (relative) amounts of starting substances as more new substances form.

For hexamethylenediamine and adipic acid:

- Show Video—*Hexamethylenediamine and Adipic Acid* (Step 1).
- Students discuss and record their responses in Table 1.8 and for Step 2.
- Lead discussion of student observations and ideas about Step 2.

2. Use your observations to answer the following questions about the starting and ending substances:

   a. What happens to the amounts of the two liquids as more nylon thread is pulled from the beaker? What is your evidence?

   The amount of each of the two liquids decreases as more nylon thread is pulled from the beaker. We can see the level of each liquid decrease in the beaker.

   b. Where do you think the new substance (nylon) is coming from? Support your claim with evidence.

   I think the nylon is somehow coming from the two starting substances. My evidence is that it looks like the scientist is pulling the nylon thread from the place where the two liquids meet and that the amount of each liquid is decreasing as the amount of nylon is increasing.

3. Watch the video of the change involving steel wool (iron) and oxygen in the air. Look closely at how the amounts of starting and ending substances are changing from the beginning to the end of the video. Record the names of the substances and your observations in Table 1.9.

Table 1.9. Steel Wool (Iron) and Air (Oxygen)

Describe what happens to the amounts of the starting substances.	
Name	What happens to the amount?
Steel Wool (Iron)	*It decreases.*
Name	What happens to the amount?
Air (Oxygen)	*It decreases.*
Describe what happens to the amounts of the ending substance.	
Name	What happens to the amount?
Rust	*It increases.*

4. Use your observations to answer the following questions about the starting and ending substances:

   a. What happens to the amount of oxygen gas in the flask as more rust forms? What is your evidence?

   The amount of oxygen gas in the flask decreases as more rust forms. We observed the balloon get pushed inside the flask. We observed that a burning splint went out (negative splint test) when it was inserted into the flask, indicating oxygen was gone.

Teacher Talk and Actions

Activity 1 (cont)

Note that without a model for thinking about how different substances can be made from starting substances, the best students can do is recognize the correlation between increasing amounts of ending substances and decreasing amounts of starting substances. So don't push for a precise answer about where the nylon comes from. It's fine for students to say that it probably, somehow, comes from some kind of interaction between the hexamethylenediamine and adipic acid. They can conclude that it must come from an interaction because nylon is only formed where the liquids touch.

Students may use the term *chemical reaction* without having a model for thinking about what is actually going on. They will have the opportunity to develop a model in Lesson 1.5.

For steel wool (iron) and air:

- Show Video—*Steel Wool (Iron) and Air With Splint Test* (Step 3).

Science Note to Teacher:
Because rust forms on the steel wool, some students may come to class thinking the products of rusting are the same substances as the reactants but with different properties. But the video should help students

- observe that there is less unrusted steel wool with longer and more exposure to air, and
- infer from splint tests that less oxygen gas is present in the flask after rusting has occurred.

b. What happens to the amount of blackish-gray (unrusted) steel wool in the flask as more rust forms? What is your evidence?

The amount of unrusted steel wool decreases as more rust forms. The rust forms on the surfaces of the steel wool pad that are exposed to air. After a short time, only the outer parts of the steel wool pad have rusted and the rest looks unrusted. After more time, and if more of the steel wool pad is exposed to air, more of the steel wool rusts.

c. Where do you think the new substance (rust) is coming from?

The rust could somehow be coming from the starting substances (steel wool and oxygen).

5. Now we will investigate the change involving baking soda and vinegar again.

   a. Carefully place 1 teaspoon of baking soda and the cup containing 25 ml of vinegar in the bag. Be careful not to let the substances mix. Smell the starting substances.

   b. Squeeze as much of the air out of the bag as possible and seal it. Then turn the cup over and mix the baking soda and vinegar thoroughly.

   c. Look closely at how the amounts of starting and ending substances are changing. Record the names of the substances and your observations in Table 1.10.

Table 1.10. Baking Soda and Vinegar

Describe what happens to the amounts of the starting substances.	
Name	What happens to the amount?
Baking Soda (Sodium bicarbonate)	*It decreases.*
Name	What happens to the amount?
Vinegar (Acetic acid)	*It decreases.*
Describe what happens to the amounts of the ending substances.	
Name	What happens to the amount?
Carbon dioxide	*It increases.*
Name	What happens to the amount?
Water	*It increases.*
Name	What happens to the amount?
Sodium acetate	*It increases.*

Teacher Talk and Actions

Activity 1 (cont)

Step 4c: Note that without a model for thinking about how different substances can be made from starting substances, the best students can do is recognize the correlation between increasing amounts of ending substances and decreasing amounts of starting substances. So don't push for a precise answer about where the rust comes from. It's fine for students to say that it probably, somehow, comes from some kind of interaction between the iron and the oxygen. They can conclude that rust probably comes from an interaction between iron and oxygen because rusting stops when the oxygen in the closed flask is used up. More rust is formed only when the container is opened so more oxygen can get in.

Students may use the term *chemical reaction* without having a model for thinking about what is actually going on. They will have the opportunity to develop a model in Lesson 1.5.

Distribute materials to teams.

Remind students of the procedure they used in Lesson 1.2:

- Place a teaspoon of baking soda in the zipper plastic bag.
- Measure 25 mL of vinegar in the measuring cup.
- Place the cup upright in the bag without spilling.
- Press air out of the bag and seal it tightly.
- Tip the cup of vinegar to mix it with baking soda.

6. Use your observations to answer the following questions about the starting and ending substances:

   a. What happens to the amount of baking soda in the bag as more carbon dioxide gas forms? What is your evidence?

   The amount of baking soda in the bag appears to decrease as more carbon dioxide gas forms. There appears to be less of the white, powdery solid.

   b. Open the bag and smell the contents. Based on smell, what do you think happens to the amount of vinegar in the bag as more carbon dioxide gas forms? What is your evidence?

   The amount of vinegar in the bag appears to decrease as more carbon dioxide gas forms. The strong smell of vinegar is nearly gone.

   c. Where do you think the new substance (carbon dioxide) is coming from?

   The carbon dioxide gas could somehow be coming from the baking soda and vinegar starting substances.

7. Write a general statement that describes how changes in the amounts of the starting and ending substances are related.

   As the amounts of the ending substances increase the amounts of the starting substances decrease. OR

   As the amounts of the starting substances decrease the amounts of the ending substances increase.

8. What does your general statement suggest about where the ending substances come from?

   The starting substances must somehow interact to form the ending substances. Our observations that a decrease in the amounts of the starting substances is always accompanied by an increase in the amounts of the ending substances provides evidence that the changes in amounts of starting and ending substances are somehow related and suggests that the ending substances come from the starting substances.

Teacher Talk and Actions
Activity 1 (cont)
Step 6c: Note that without a model for thinking about how different substances can be made from starting substances, the best students can do is recognize the correlation between increasing amounts of ending substances and decreasing amounts of starting substances. So don't push for a precise answer about where the carbon dioxide comes from. It's fine for students to say that it probably, somehow, comes from some kind of interaction between the baking soda and vinegar. They can conclude that it must come from an interaction because bubbling doesn't start until the baking soda and vinegar are mixed. Students may use the term *chemical reaction* without having a model for thinking about what is actually going on. They will have the opportunity to develop a model in Lesson 1.5.

9. If we had the right equipment, we could actually measure the amounts of the starting and ending substances at different time points during the change involving hexamethylenediamine and adipic acid. The first two graphs below show how the amount of hexamethylenediamine and the amount of nylon changed, starting from the time we mixed the starting substances (start) to the time all the starting substances were gone (end). Draw a line on each of the two bottom graphs to show how the amounts of adipic acid and water changed over time.

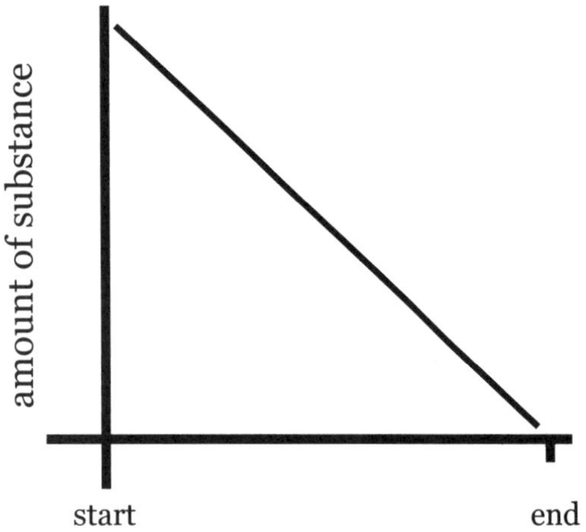

Hexamethylenediamine

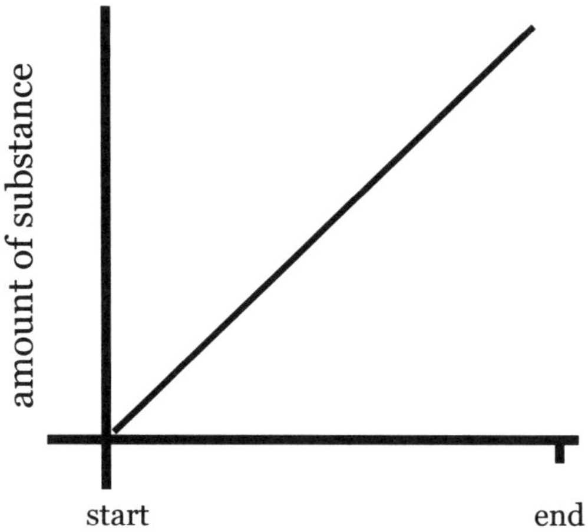

Nylon

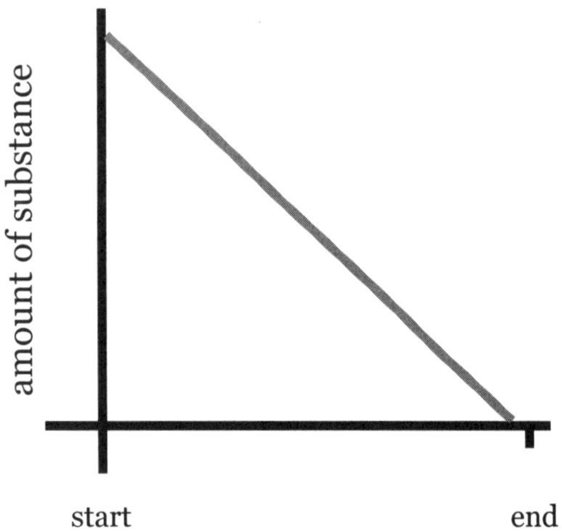

Adipic Acid

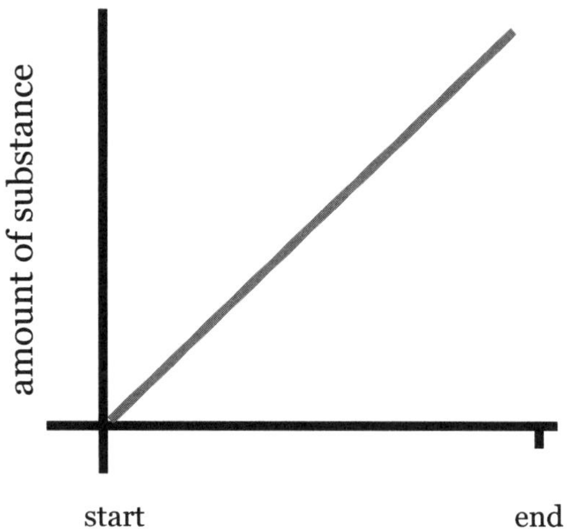

Water

SE p. 29

Teacher Talk and Actions
Activity 1 (cont)
Step 9: To check for students' understanding of the graphs, ask the following questions, "Based on the graphs, what substances are present at the start of the reaction? What substances are present at the end of the reaction?"

Science Ideas

Science ideas are accepted principles or generalizations about how the world works based on a wide range of observations and data collected and confirmed by scientists. Because these science ideas are consistent with the available evidence, you are justified in applying science ideas to new observations and data.

Lesson 1.3 was intended to help you understand an important idea about how to tell when new substances form and where they might come from. Read the idea below. Look back through Lesson 1.3. In the space provided after the science idea, give evidence that supports the idea.

Science Idea #2: Changes during which new substances form are called chemical reactions. The correlation of increasing amounts of ending substances with decreasing amounts of the starting substances provides evidence that the new substances result from an interaction between the starting substances.

Evidence:

When steel wool (iron) and air (oxygen) were mixed, we saw that only a little rust formed in the sealed flask but that more rust formed when the flask was opened to the air. We also saw that the oxygen in the sealed flask was gone after the little bit of rust had formed. So the iron and oxygen interacted to form rust. We can tell that rust is a new substance because it has different properties (reddish-brown solid) from iron (grayish metallic solid) and oxygen (colorless, odorless gas that gives a positive splint test).

Teacher Talk and Actions	
Science Ideas	**(10 min)**

Whole class:

Tell students: You have used your own observations and scientists' data about the characteristic properties of substances to develop an important idea about a particular kind of change in matter, called chemical reactions.

Read the science idea(s) aloud to the class. Add to the list of charted science ideas. Give students time to write down at least one piece of evidence that illustrates the idea.

Remind students that the science ideas will be used in constructing scientific explanations, such as in Activity 2.

Activity 2: Constructing Scientific Explanations—Part II

In Lesson 1.2, you examined Eva's explanation for her claim that a new substance formed when hexamethylenediamine and adipic acid were mixed and decided that her explanation answered all of the Explanation Quality Questions. In this activity, you will use the Explanation Quality Questions to evaluate an explanation and improve it.

Procedures and Questions

1. Read Jeremy's explanation to the question, Did a new substance form when baking soda and vinegar were mixed, and if so, where did it come from?

   Carbon dioxide was produced from a chemical reaction between baking soda and vinegar. If a new substance is produced, then a chemical reaction has occurred (Science Idea #2) and the properties changed when baking soda and vinegar were mixed. My evidence that carbon dioxide was produced is that it turned limewater cloudy.

2. Now answer the Explanation Quality Questions for Jeremy's explanation. You can refer to Table 1.6 in Lesson 1.2 to refresh your memory about what each part is.

   - Does the **claim** respond to the question?

   The claim doesn't completely respond to the question. It states that carbon dioxide was produced by a chemical reaction between baking soda and vinegar, but it doesn't state that a new substance is produced

   - Are the **science ideas** that are listed relevant to the claim?

   The idea that if a new substance (with different properties from the starting substances) is produced, then a chemical reaction has occurred (Science Idea #2) is relevant to the claim.

   - Does the **evidence** support the claim?

   Jeremy's explanation does not provide evidence because properties of a substance don't change during a chemical reaction. Instead, his explanation should have stated that a substance was produced that had different properties (a gas at room temperature) than the starting substances (a solid and a liquid at room temperature).

   - Are the **science ideas** used to show why the evidence supports the claim?

   No. Using Science Idea #2 would require a statement like this: Since the gas produced has different properties from the starting substances, it is a different substance. And since a different substance was produced, a chemical reaction occurred.

Teacher Talk and Actions	
Activity 2	**(25 min)**
Evidence Explanation Quality Question: The statement "the properties changed" is vague. Does Jeremy think that the properties of a substance can change? If so, he is incorrect. Make sure students understand that they should be precise in their wording.	

3. Now it is your turn to write an explanation to the question, Did a new substance form when baking soda and vinegar were mixed, and if so, where did it come from?

   a. Record your notes for each part of the explanation in the table below:

Question	Did a new substance form when baking soda and vinegar were mixed, and if so, where did it come from?
Claim	Yes, a new substances formed when baking soda and vinegar were mixed.
Science Ideas	Changes during which starting substances interact to form new substances are called chemical reactions. The ending substances can be recognized as new substances because they have different properties than the starting substances (Science Idea #2).
Evidence	Baking soda is a white solid and vinegar is a clear, colorless, smelly liquid. When they were mixed, both the white powder and the smelly liquid disappeared, but a gas that gave a positive limewater test appeared that wasn't initially present.

   b. Use the information from the table to write your explanation.

A new substance formed when baking soda and vinegar were mixed. Changes during which starting substances interact to form new substances are called chemical reactions. The ending substances can be recognized as new substances because they have different properties from the starting substances (Science Idea #2). Baking soda is a white solid and vinegar is a clear, colorless, smelly liquid. When they were mixed, both the white powder and the smelly liquid disappeared, but a gas appeared that wasn't initially present. The gas gave a positive limewater test. Since the ending substance had different properties from the starting substance, a new substance was produced during a chemical reaction between baking soda and vinegar.

Teacher Talk and Actions
Activity 2 (cont)
(Intentionally left blank)

Pulling It Together

Work on your own to answer these questions. Be prepared for a class discussion.

1. After observing the three changes again, how would you answer the Key Question, **As a new substance forms, what happens to the starting substances?**

The starting substances interact to form the ending substances. This is why the ending substances are only produced when the starting substances interact (mix) and why the amounts of the starting substances decrease as the amounts of the ending substances increase.

2. A friend tells you that the burning of a candle involves a chemical reaction, during which oxygen in the air interacts with the candle to produce carbon dioxide and water. Your friend provides evidence listed in the table below. For each statement, indicate what it provides evidence for.

Observation	What It Provides Evidence for
A large beaker is placed over a burning candle. Before the beaker is put over the candle, a burning splint stays lit. However, after the beaker is placed over the candle and the flame on the candle goes out, the air in the beaker puts out the burning splint.	*Oxygen is used, because if oxygen was present, a burning splint would continue to burn.*
A large beaker is placed over a burning candle. Before the beaker is put over the candle, the air in the beaker does not turn limewater cloudy. However, after the beaker is placed over the candle and the flame on the candle goes out, the air in the beaker turns limewater cloudy.	*Carbon dioxide is produced, because now the air inside the beaker turns limewater cloudy—which is an indication that carbon dioxide is present.*
When a large beaker is placed over a burning candle, as the amount of the candle decreases, the amount of oxygen in the beaker also decreases. Meanwhile, the amount of carbon dioxide and water in the beaker increases.	*The correlation of decreasing reactants and increasing products gives evidence that the oxygen and candle are used to create the carbon dioxide and water.*

Teacher Talk and Actions

Pulling It Together (15 min)

Closure and Link (2 min)

Whole class:

Summarize the lesson:

We have evidence that new substances can form during chemical reactions, and that these new substances are produced when the starting substances interact. We inferred this because (a) the ending substances didn't form until the starting substances were mixed and (b) the amount of the starting substances decreases as the amount of the ending substances increases.

We could recognize them as new substances because they had different characteristic properties from the starting substances.

Link to the next lesson:

Elicit a few students' ideas to this question: If the new substances come from the starting substances, why do they have such different properties?

Lesson Guide
Chapter 1, Lesson 1.4
Using Models to Think About Atoms and Molecules

Focus: In this lesson, students examine models of starting and ending substances to visualize the parts of substances (atoms and molecules) they cannot see. Students learn that all substances are made up of atoms, but different substances are made up of different atoms or different combinations of atoms (called molecules).

Key Question: Why do the substances produced in a chemical reaction have different properties from the starting substances?

Target Idea(s) Addressed

- A molecule is made up of two or more atoms connected together in a specific arrangement.
- Each substance is made up of a single type of atom or molecule.
- The properties of a substance are determined by the type, number, and arrangement of atoms that the substance is made up of.
- The starting and ending substances of chemical reactions have different properties because the molecules that make them up have different arrangements of atoms.

Science Practice(s) Addressed

- Analyze and interpret data to provide evidence for phenomena.
- Use multiple models to represent phenomena.

Materials

Activity 1: *Card Pack 3,* Copymaster: Model Key

Activity 2: *Card Packs 1, 2,* and *3*; Blank Change Charts

Advance Preparation

Organize materials in advance so that each set of partners or table teams has materials when class begins.

Phenomena, Data, or Models	Intended Observations	Purpose	Rationale or Notes
Activity 1 Cards representing atomic/molecular composition of substances with a variety of models (LEGO bricks, ball-and-stick models, space-filling models, chemical symbols and formulas, structural formulas)	Different models can be used to represent the atomic/molecular composition of water. All models of water molecules have the same atomic composition (two hydrogen atoms and one oxygen atom). Water, iron oxide, acetic acid, iron, carbon dioxide, sodium bicarbonate, and oxygen are all made up of atoms, and different substances are made up of different atoms or combinations of atoms (molecules).	Students interpret a variety of atomic/molecular representations of substances and recognize when different representations are of the same molecule, and when the representations are of different molecules.	This activity uses models to illustrate that all matter, even gases, is particulate. The activity also provides opportunities to challenge students' confusion about the difference between atoms and molecules.
Activity 2 Change charts with data cards of starting and ending substances (that now include atomic/molecular composition) and used to represent reactants and products of chemical reactions.	Both the starting and ending substances of the iron rusting reaction are made up of Fe and O atoms. Both the starting and ending substances of the baking soda + vinegar reaction are made up of H, C, O, and Na atoms. Both the starting and ending substances of the nylon forming reaction are made up of H, C, O, and N atoms.	Students have the opportunity to generalize across observations of models of starting and ending substances that molecules are made up of atoms before reading Science Idea #3. Students have the opportunity to generalize across observations of models and data that when substances have different atomic compositions they have different properties before reading Science Idea #4.	This activity uses models to help students think about why different substances have different properties. The activity also provides an opportunity to challenge students' misconception that the properties of substances change during a chemical reaction, rather than the correct idea that new substances are produced that have different properties. The properties of a substance are part of the identity of the substance and do not change.
Pulling It Together Data about the atomic composition of starting and ending substance	When white crystals (formed from chemical reaction between grey metallic solid and yellow gas) are ground to a white powder, both the crystals and the powder have the same atomic composition (NaCl).	Students have the opportunity to explain why a change is not a chemical reaction, based on both starting and ending substance having the same formula (atomic composition).	This activity uses models to help explain why two substances are the same (and also why texture is not a characteristic property). The activity also challenges a common misconception that all changes involve chemical reactions.

Lesson 1.4—Using Models to Think About Atoms and Molecules

What do we know and what are we trying to find out?

So far, we've learned that every substance has a unique set of properties that can be observed and measured. We also know that during chemical reactions, substances form that have different properties from the starting substances. How can this happen? If a substance can't change its characteristic properties, where do the new substances come from? Why do different substances have different properties?

To answer these questions, we need to know more about the matter that substances are made of. You may have heard that all matter—that is, every single substance—is made up of atoms. But if all substances are made up of atoms, how can two substances be different? If substances are made up of the same atoms, will they always be the same? Because atoms are too small to be seen, in this lesson you will use models to help you think about these things.

Answer the Key Question to the best of your knowledge. Be prepared to share your ideas with the class.

Key Question: Why do the substances produced in a chemical reaction have different properties from the starting substances?

Maybe the starting substances and ending substances are made up of different things.

Teacher Talk and Actions
What do we know and what are we trying to find out? (3 min)
Whole class: Read text under "What do we know and what are we trying to find out?" aloud to the class.
Key Question (5 min)
Individuals: Students respond to the Key Question: **Why do the substances produced in a chemical reaction have different properties from the starting substances?** Whole class: Lead discussion. Invite students to share ideas about the Key Question. The goal is primarily eliciting and probing student ideas here. When students share their answers, do not indicate if their answers are right or wrong. However, do not let students assume by your silence that they were correct. Make it clear that students are expressing ideas, but those ideas may change based on what they will see in the lessons. (Remind students of this regularly.) Challenge ideas that are inconsistent with learning from previous lessons. Students may confuse elements, compounds, atoms, and molecules. Do not correct them at this point.

Activity 1: Comparing Models of Different Substances

Materials
For each team of students
Card Pack 3
Model Key

In this activity, you will look at cards that show different kinds of models of the molecules of different substances. The cards show chemical formulas, structural formulas, LEGO models, ball-and-stick models, and space-filling models, which can all be used to represent the molecules of each substance. You will analyze and interpret these models to help you describe what different substances are made up of and to help you explain why different substances have different properties.

Procedures and Questions

1. Find the Water card in *Card Pack 3* and study the different models on it.

   Use the *Model Key* to help you make sense of the models you are looking at.

2. Answer the questions below based on your observations of the models for a water molecule (see Table 1.11, p. 36).

   a. How does each model represent a water molecule?

   All of the water molecule models have a way of showing that there are two hydrogen atoms and one oxygen atom, and that the three atoms are connected to one another.

   b. How is an atom different from a molecule?

   Models show that a water molecule is made up of atoms and atoms make up a molecule.

   c. How would you explain to a friend what water is made up of? Be as descriptive as possible.

   Water is a substance that is made up of water molecules. Each water molecule is composed of two hydrogen atoms connected to a single oxygen atom.

Teacher Talk and Actions
Activity 1 **(20 min)**
Whole class: Read introductory text. Emphasize the importance of different kinds of models in helping us think about atoms and molecules of substances. Distribute materials to teams. Small group: Students complete Steps 1 and 2a–c. Students compare and contrast different types of models of a water molecule. Whole class: Lead discussion. Invite students to share ideas about Step 2b and Step 2c.

Table 1.11. Models of a Water Molecule

Model Type	Models of a Water Molecule
LEGO model	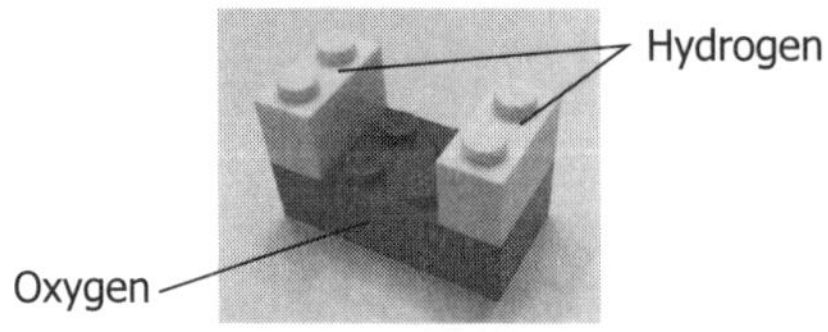
Ball-and-stick model	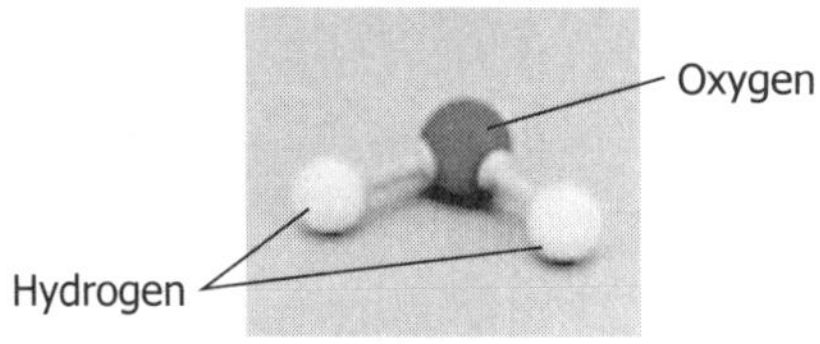
Space-filling model	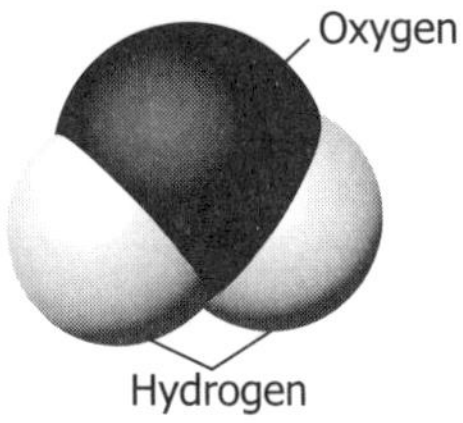
Chemical formula	$\mathbf{H_2O}$
Structural formula	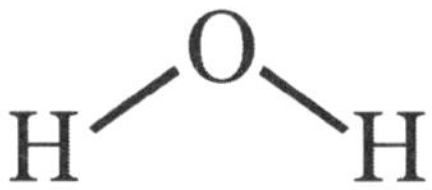

3. Participate in a class discussion about the similarities and differences between the models and between the models and the real thing that they represent.

4. Spread the rest of the cards from *Card Pack 3* face up on your desk or table.

5. Study the different models on each card. Think about what each substance is made up of:

    a. Does the model show that this substance is an atom or a molecule?

    b. If it is a molecule, then what type(s) of atoms and how many of each make up the molecule?

Teacher Talk and Actions
Activity 1 (cont)
Whole class: Lead a discussion about models, what they represent, and similarities and differences between models and what they represent. Be sure students realize that all the models show the same atomic composition of the water molecule. <u>Science Note to Teacher:</u> • Iron is a metallic solid, made up of a single type of atom, packed together in orderly arrays. For simplicity, it is represented as Fe. • Iron oxide is an ionic solid, made up of iron and oxygen atoms. For simplicity, it is represented as a molecule. • Nylon is a polymer made up of thousands of alternating monomers of hexamethylenediamine and adipic acid that have undergone chemical reactions as they link together. For simplicity, it is represented as a dimer. Students will learn more about nylon polymerization in Lesson 1.6.

6. Fill in Table 1.12. What is each substance made up of? Two examples have been done for you.

Table 1.12. What Are Substances Made Up Of?

Substance Name	Observable Properties	Atom or Molecule	Composition
Iron oxide (rust)	*Reddish orange flaky solid*	*Molecule*	*3 oxygen atoms and 2 iron atoms*
Acetic acid (vinegar)	*Colorless smelly liquid*	*Molecule*	*2 carbon atoms, 4 hydrogen atoms, and 2 oxygen atoms*
Iron (steel wool)	*Gray shiny metal solid*	*Atom*	*1 iron atom*
Carbon dioxide gas	*Colorless gas*	*Molecule*	*1 carbon atom and 2 oxygen atoms*
Sodium bicarbonate (baking soda)	*Colorless smelly liquid*	*Molecule*	*1 sodium atom, 1 hydrogen atom, 1 carbon atom, and 3 oxygen atoms*
Oxygen gas	*Colorless gas*	*Molecule*	*2 oxygen atoms*
Water	*Colorless liquid*	*Molecule*	*1 oxygen atom and 2 hydrogen atoms*

Teacher Talk and Actions
Activity 1 (cont)
Partners/teams: Fill in Table 1.12. Seek feedback from other teams on the accuracy of information shared. Students should identify new information provided on the substance cards: • Atom composition (type of atoms and number of each type of atom) • Chemical formulas (another way of representing the type and number of each type) • Space-filling model (also shows which atoms are linked) Emphasize ideas which may not be evident to students from the cards alone: • The cards illustrate important ideas that apply to all substances. • Molecules of the same substance have the same composition (same types of atoms and same number of each type) and same arrangement of atoms. • Molecules of one substance are alike throughout any sample of that substance, but they are different from the molecules of another substance. • Chemical formulas are based on evidence from experiments.

Activity 2: Comparing Starting and Ending Substances

Materials
For each team of students
Card Pack 1 (green outline)
Card Pack 2 (yellow outline)
Card Pack 3
Model Key
Blank Change Charts

In this activity, you will learn more about the atoms and molecules that make up the starting and ending substances involved in the three chemical reactions that we have been exploring. You will use information on substance cards to think about why the new substances formed during a chemical reaction have properties that are different from those of the starting substances.

Procedures and Questions

1. Find the *Change Chart* for Steel Wool (Iron) and Air (Oxygen). Look in *Card Pack 1* for the two starting substance cards, and place them on the left (starting substance) side of the chart. Then look in *Card Pack 2* for the ending substance, and place that card on the right (ending substance). You already know that the ending substance (iron oxide) has different properties from the starting substances (iron and oxygen). Let's see if we can use the models of the substances in *Card Pack 3* to figure out why.

2. First, using *Card Pack 3*, look at the types of atoms making up the molecules of the starting substances and the ending substances, and record the information in the first row of Table 1.13.

3. Look at the ball-and-stick models or the structural formulas for the starting and ending substances, and describe how the atoms are arranged in the second row of Table 1.13.

Table 1.13. Steel Wool (Iron) and Air (Oxygen)

	Starting Substances	Ending Substances
Types of Atoms	Iron Oxygen	Iron Oxygen
Arrangement of Atoms	In iron, Fe atoms are not connected to other atoms. In oxygen gas, each O atom is connected to one other O atom.	In iron oxide, the Fe and O atoms are connected to each other—2 Fe atoms and 3 O atoms make up iron oxide.

Teacher Talk and Actions	
Activity 2	**(10 min)**
Whole class: Read introductory text. Distribute materials to teams.	

4. How do the types of atoms in the starting substances compare with the types of atoms in the ending substances?

   Both the starting and ending substances are made up of the same types of atoms.

5. Compare the arrangement of the atoms in the starting and ending substances.

   The arrangement of Fe and O atoms is different in the starting and ending substances. For example, in the starting substances two O atoms are connected to each other and Fe atoms are not connected to other atoms. However, in the ending substance each Fe atom is connected to two O atoms and one of the O atoms is connected to two Fe atoms.

6. Based on the information in Table 1.13, why do you think iron oxide might have different properties from iron and oxygen?

   Even though the iron oxide is made up of the same types of atoms as the iron and oxygen gas, the atoms are arranged differently in the ending substance. Perhaps the reason iron oxide has different properties is because the Fe and O atoms are arranged differently.

7. Find the *Change Chart* for Hexamethylenediamine and Adipic Acid. Look in *Card Pack 1* for the two starting substance cards, and place them on the left (starting substance) side of the chart. Then look in *Card Pack 2* for the ending substances, and place those cards on the right (ending substance).

8. Then, using *Card Pack 3,* look at the types of atoms making up the molecules of the starting substances and the ending substances, and record the information in the first row of Table 1.14.

9. Look at the ball-and-stick models or the structural formulas for the starting and ending substances, and describe how the atoms are arranged in the second row of Table 1.14.

Table 1.14. Hexamethylenediamine and Adipic Acid

	Starting Substances	Ending Substances
Types of Atoms	*Nitrogen, Carbon, Hydrogen, Oxygen*	*Nitrogen, Carbon, Hydrogen, Oxygen*
Arrangement of Atoms	*In hexamethylenediamine, carbon atoms are connected to carbon atoms, hydrogen atoms, and sometimes nitrogen atoms.* *In adipic acid, carbon atoms are connected to other carbon atoms, hydrogen atoms, and sometimes oxygen atoms.*	*In water, 2 hydrogen atoms are connected to an oxygen atom.* *In nylon, there is a new connection between nitrogen from hexamethylenediamine and the carbon that used to be connected to oxygen in adipic acid.*

Teacher Talk and Actions
Activity 2 (cont)
(Intentionally left blank)

10. How do the types of atoms in the starting substances compare with the types of atoms in the ending substances?

Both the starting and ending substances are made up of the same types of atoms.

11. Compare the arrangement of the atoms in the starting and ending substances.

The arrangement of C, O, N, and H atoms is different in the starting and ending substances. For example, the N of hexamethylenediamine is connected to the C from adipic acid in nylon. Also, the O atom in water is no longer connected to a C atom.

12. Based on the information in Table 1.14, why do you think nylon and water might have different properties from heyxamethylenediamene and adipic acid?

Even though the nylon and water are made of the same types of atoms as hexamethylenediamine and adipic acid, the atoms are arranged differently. Perhaps the reason nylon and water have different properties is because the atoms are arranged differently.

13. Find the *Change Chart* for Baking Soda and Vinegar. Look in *Card Pack 1* for the two starting substance cards, and place them on the left (starting substance) side of the chart. Then look in *Card Pack 2* for the ending substances, and place those cards on the right (ending substance).

14. Look at the types of atoms making up the molecules of the starting substances and the ending substances, and record the information in the first row of Table 1.15.

15. Look at the ball-and-stick models or the structural formulas for the starting and ending substances, and describe how the atoms are arranged in the second row of Table 1.15.

Table 1.15. Baking Soda and Vinegar

	Starting Substances	**Ending Substances**
Types of Atoms	*Sodium* *Carbon* *Oxygen* *Hydrogen*	*Sodium* *Carbon* *Oxygen* *Hydrogen*
Arrangement of Atoms	*In baking soda, there are sodium, carbon, oxygen, and hydrogen atoms, while there are only carbon, oxygen, and hydrogen atoms in vinegar.*	*In carbon dioxide, only carbon atoms and oxygen atoms are connected. Water has oxygen and hydrogen atoms that are connected.* *Sodium acetate has all the types of atoms.*

SE p. 40

Teacher Talk and Actions
Activity 2 (cont)
(Intentionally left blank)

16. How do the types of atoms in the starting substances compare with the types of atoms in the ending substances?

Both the starting and ending substances are made up of the same types of atoms.

17. Compare the arrangement of the atoms in the starting and ending substances.

The arrangement of C, O, Na, and H atoms is different in the starting and ending substances. For example, the C of baking soda is connected to an H atom instead of an Na atom in the ending substances. Also, the O atom in water is not connected to a C atom now.

18. Based in the information in Table 1.15, why do you think sodium acetate, water, and carbon dioxide might have different properties from baking soda and vinegar?

Even though the sodium acetate, water and carbon dioxide are made of the same types of atoms as baking soda and vinegar, the atoms are arranged differently. Perhaps the reason the ending substances have different properties is because the atoms are arranged differently from the starting substances.

19. Based on what you have written for each of the three reactions, write a general statement that explains why the ending substances of a chemical reaction have different properties from the starting substances.

Even though molecules of the ending substances of a chemical reaction are made up of the same types of atoms as the starting substances, the atoms are arranged differently. Maybe differences in properties are due to differences in arrangements of atoms.

Teacher Talk and Actions
Activity 2 (cont)
(Intentionally left blank)

Science Ideas

Activities 1 and 2 were intended to help you understand what substances are made of and give you ideas about why different substances have different properties. Read the science ideas listed below. Remember that because science ideas are consistent with a wide range of relevant evidence, you can use them to explain observations and data about other substances. Look back through Lesson 1.4 and give at least one piece of evidence that supports each science idea.

Science Idea #3: A molecule is made up of two or more atoms connected together in a specific arrangement.

Evidence:

A water molecule is made up of two hydrogen atoms and one oxygen atom that is connected to both hydrogen atoms.

Science Idea #4: Each substance is made up of a single type of atom or molecule. The properties of a substance are determined by the type, number, and arrangement of atoms that it is made up of. Because no two substances are made up of the same arrangement of atoms, no two substances have the same set of properties.

Evidence:

Our observations showed that oxygen gas has different properties from carbon dioxide gas. While both are clear, colorless gases, oxygen gives a positive splint test (carbon dioxide does not) and carbon dioxide gives a positive limewater test (oxygen does not). Data on the substance cards showed that oxygen has a melting point of -78°C, whereas carbon dioxide has a melting point of -219°C.

And the atomic composition of oxygen is O_2, whereas the atomic composition of carbon dioxide is CO_2. According to Science Idea #4, the reason oxygen gas and carbon dioxide gas have different properties, such as melting point and density, is because oxygen gas is made up of O_2 molecules whereas carbon dioxide gas is made up of CO_2 molecules.

Teacher Talk and Actions	
Science Ideas	**(5 min)**

Whole class:

Tell students: You have used models that represent the atoms and molecules of substances to help you visualize important science ideas.

Read the science idea(s) aloud to students. Add Science Idea #3 and Science Idea #4 to the list of charted science ideas. Give students time to write down at least one piece of evidence that illustrates the idea.

Remind students that the science ideas will be used in constructing scientific explanations later on.

Pulling It Together

Work on your own to answer these questions. Be prepared for a class discussion.

1. After studying the models, how would you answer the Key Question, **Why do the substances produced in a chemical reaction have different properties from the starting substances?**

The models helped us see that different substances are made up of different atoms or different combinations of atoms linked together to form molecules. Models helped us visualize the specific types and number of atoms that made up the molecules of a particular substance. Some models also helped us see how the atoms are joined together. The specific types of atoms, number of atoms, and how the atoms are connected in molecules are different for different substances and might be the reason why different substances have different properties.

2. Recall the scientist who collected the white crystals that resulted from mixing a yellow gas and grayish metal together. She determined that the chemical formula for the crystals was NaCl. She then took the white crystals from the container and ground them down into a very fine powder. The texture of the powder was different from the texture of the crystals. But when the scientist measured the conductivity and solubility of the powder, she found them to be the same as the conductivity and solubility of the crystals. She then determined that the chemical formula for the powder was NaCl.

   a. Is a new substance made when the crystals are ground down into a powder? Use the table below to make notes and then write a valid explanation.

Question	Is a new substance made when the crystals are ground into a powder?
Claim	*No. No new substances form when the white crystals are ground into a powder.*
Science Ideas	*Every substance has a unique set of characteristic properties, such as color, odor, density, melting point, conductivity and solubility, and how it behaves (such as in limewater and glowing splint tests). The properties of substances can be observed or measured and used to decide if two substances are the same or different (Science Idea #1).*
Evidence	*The white crystals have the same molecular formula as the white powder that forms after grinding.*

Teacher Talk and Actions

Pulling It Together **(5 min)**

Individuals:

Students answer Pulling It Together questions on their own, in detail.

Direct students to look back at their initial ideas about the Key Question and consider how their ideas have changed.

Whole class:

Lead a class discussion. Invite students to share ideas about Pulling It Together Question 2.

Explanation:

No, a new substance is not formed when the white crystals are ground into a powder. Every substance has a unique set of characteristic properties, such as color, odor, density, melting point, conductivity and solubility, and how it behaves (such as in limewater and glowing splint tests). The properties of substances can be observed or measured and used to decide if two substances are the same or different (Science Idea #1). The white powder has the same set of properties as the white crystals, including the same molecular formula. (Texture is not a characteristic property.) Since both substances have the same set of characteristic properties, they are the same substance.

3. In this lesson, you've seen how different kinds of models can be used to show the atoms and molecules that make up the starting and ending substances involved in chemical reactions. You've also used the models to think about why the starting and ending substances have different properties, even though they are both made of the same types of atoms. How do you think you can use these same models to help you figure out why new substances are produced in chemical reactions?

Maybe we could build models of the molecules of the starting substances and see whether we could use those bricks to build molecules of the ending substances. We saw that models of molecules of both starting and ending substances of the 3 chemical reactions were made up of the same types of atoms. I'm not sure if molecules of starting and ending substances are made up of the same number of each type of atom, but we could model the reaction to see if it works.

Teacher Talk and Actions
Closure and Link **(2 min)**
Whole class: Summarize the lesson: We used models to reason about the smallest parts of substances. We learned that different substances have different properties because of the atoms or combination of atoms (molecules) they are made up of. Link to next lesson: In the next lesson, we will use LEGO models to represent what happens to the atoms that make up starting substances to make the molecules of new substances.

Lesson Guide
Chapter 1, Lesson 1.5
Using Models to Represent Chemical Reactions

Focus: The lesson uses LEGO models to show that during chemical reactions, atoms of starting substances rearrange and reconnect in different combinations to form molecules of new substances. The lesson confronts common ideas that atoms are created or destroyed or turned into new atoms during chemical reactions.

Key Question: How are the molecules of the ending substances made?

Target Idea(s) Addressed

- During chemical reactions, atoms that make up molecules of the starting substances (called reactants) separate from one another and connect in different ways to form the molecules of the ending substances (called products).

Science Practice(s) Addressed

- Use a model to describe phenomena.
- Construct an explanation supported by empirical evidence and scientific reasoning to support a claim.

Materials

Activity 1: LEGO model kit, Copymasters: *Chemical Reaction Mats*: Steel Wool (Iron) and Air (Oxygen), Baking Soda and Vinegar, *Model Key*

Advance Preparation

Organize materials in advance so that each set of partners or table teams has materials when class begins.

Phenomena, Data, or Models	Intended Observations	Purpose	Rationale or Notes
Activity 1 LEGO models representing two chemical reactions: • Rust forms when steel wool (iron) and air (oxygen) are mixed. • Carbon dioxide, water, and sodium acetate form when baking soda and vinegar are mixed.	Bricks of the models of iron and oxygen (4 Fe and 3 O_2) are disconnected and connected differently to make models of ending substances (2 Fe_2O_3), with no bricks left over and no additional bricks needed. Bricks of the models of baking soda ($NaHCO_3$) and vinegar ($HC_2H_3O_2$) are disconnected and connected differently to make models of ending substances (CO_2, H_2O, and $NaC_2H_3O_2$), with no bricks left over and no additional bricks needed. All the atoms making up a molecule don't necessarily disconnect during chemical reactions. Sometimes small groups of bricks from the models of starting substances stay together with only a few bricks connecting with different bricks to make models of ending substances.	Students visualize the rearrangement of atoms (and forecast the conservation of atoms) of starting substances to account for new substances that form during chemical reactions. Students have the opportunity to generalize across their observations before reading Science Idea #5.	The modeling activity challenges common misconceptions that atoms are created, destroyed, or turned into new atoms during chemical reactions. The modeling activity accurately represents the stoichiometric (lowest whole number) relationships between reactant and product molecules, but it is not an explicit focus of the activity.
Space-filling models representing methane gas and oxygen gas and possible ending substances that could form if they react	Molecules of starting substances (methane and oxygen) are collectively made up of H, C, and O atoms. Water, carbon dioxide, and carbon monoxide are collectively made up of combinations of H, C, and/or O atoms. Ammonia is made up of N and H atoms.	Students use space-filling models of molecules to predict possible ending substances given the atom composition of starting substances.	

Phenomena, Data, or Models	Intended Observations	Purpose	Rationale or Notes
Activity 2 Carbon dioxide (CO_2), sodium acetate ($NaC_2H_3O_2$), and water (H_2O) form when baking soda ($NaHCO_3$) and vinegar ($HC_2H_3O_2$) react. Rust (Fe_2O_3) forms when iron (Fe) and oxygen (O_2) react.	LEGO models showing that it is possible to rearrange atoms of molecules of reactants to form molecules of products can be used to support the claim that actual atoms of baking soda and vinegar rearrange to form carbon dioxide, water, and sodium acetate is reasonable. LEGO models showing that it is possible to rearrange atoms of molecules of reactants to form molecules of products can be used to support the claim that actual atoms of iron and oxygen rearrange to form iron oxide is reasonable.	Students observe the use of models in explaining how the ending substances might form from the starting substances in the case of the baking soda + vinegar reaction. Students have the opportunity to practice using models in explaining how the ending substances might form from the starting substances in the case of the iron rusting reaction.	Optional: To emphasize that the models only suggest how something might work but don't prove that it actually does work that way, ask different groups of students to write down the sequence of disconnecting and reconnecting bricks. Different groups will likely have different sequences even though both groups started with the same LEGO models of reactants and ended with the same LEGO models of products.
Pulling It Together LEGO models show that hydrogen peroxide molecules react to form oxygen and water molecules LEGO models show that the molecules of gas inside bubbles that form when water is heated on stove are the same as the molecules making up the unheated water	LEGO models show that H and O atoms are connected in different ways while representing hydrogen peroxide than when representing oxygen gas and water (therefore a chemical reaction has occurred). LEGO models show that H and O atoms are NOT connected in different ways when water is heated on a stove (therefore a chemical reaction has not occurred).	Students use models of the atoms and molecules of the starting and ending substances to decide whether or not a chemical reaction has occurred and to justify their decision based on science ideas and models. An ideal explanation would include (a) one or more different molecules formed/did not form using the same atoms as those that made up the starting substances, and (b) the formation of different molecules from the atoms of reactant molecules is a criterion for a chemical reaction. Pulling It Together Questions 3 and 4 can be used as embedded assessments—rubrics are provided in the online resources.	The heating water challenges the common student misconceptions that (a) all changes in materials involve chemical reactions and (b) the formation of a gas always indicates a chemical reaction has occurred. (If the water vapor was cooled to room temperature it would be a liquid, not a gas.)

Lesson 1.5—Using Models to Represent Chemical Reactions

What do we know and what are we trying to find out?

An important idea from our previous lessons was that different substances have different properties. This is because every substance is made up of its own unique combination and arrangement of atoms, which gives each substance its own unique set of properties.

So far, we have been investigating three chemical reactions:

- Steel Wool (Iron) and Air (Oxygen)
- Hexamethylenediamine and Adipic Acid
- Baking Soda and Vinegar

As we have learned, during chemical reactions, one or more substances "react" with each other and form one or more different substances. How does this happen?

In Lessons 1.5 and 1.6, you will explore what happens to the atoms that make up the molecules of the starting substances when new substances are formed during chemical reactions. You will use models of the molecules to help guide your thinking.

Answer the Key Question to the best of your knowledge. Be prepared to share your ideas with the class.

Key Question: How are the molecules of the ending substances made?

We know that molecules of new substances are made from the same kinds of atoms as starting substances. Maybe the molecules of new substances are made by regrouping the atoms of the molecules of starting substances.

Teacher Talk and Actions

What do we know and what are we trying to find out? (3 min)

Whole class:

Read text under "What do we know and what are we trying to find out?" aloud to the class.

Key Question (5 min)

Individuals:

Students respond to the Key Question: **How are the molecules of the ending substances made?**

Whole class:

Lead discussion. Invite students to share ideas about the Key Question.

The goal is primarily eliciting and probing student ideas here.

When students share their answers, do not indicate if their answers are right or wrong. However, do not let students assume by your silence that they were correct. Make it clear that students are expressing ideas, but those ideas may change based on what they will see in the lessons. (Remind students of this regularly.)

Challenge ideas that are inconsistent with learning from previous lessons.

Activity 1: Representing Chemical Reactions With Models

Materials
For each team of students
Chemical Reaction Mats: Steel Wool (Iron) and Air (Oxygen), Baking Soda and Vinegar
LEGO model kit
Model Key

In this activity, you will use LEGO bricks to represent (or to model) what is happening to the atoms and molecules in two of the three chemical reactions we have been studying. You will use the *Chemical Reaction Mats* to guide you.

The *Chemical Reaction Mats* show the LEGO models of starting substances on one side (green side), and LEGO models of the ending substances on the other (yellow side). Each LEGO brick represents an atom. You can use the information on each mat to write a word equation for that chemical reaction. The general form of a word equation is shown below:

Starting substances ⟶ **Ending substances**

The arrow means "react to form." When you flip from the green side to the yellow side of the *Chemical Reaction Mat*, you are representing "react to form."

Procedures and Questions

1. Using the *Steel Wool (Iron) and Air (Oxygen) Chemical Reaction Mat* and LEGO model kit, do the following:
    a. Use LEGO bricks to build models of the molecules of the starting substances shown on the green side of the mat.
    b. Once you have built the models of the starting molecules exactly as they are shown on the mat, put all unused LEGO bricks back in the model kit and close the kit.
    c. Look closely at the starting molecule models you built and the ending molecule models shown on the yellow side of the mat. Think about how the models of starting molecules and ending molecules are similar to and different from each other.
    d. Using only the LEGO bricks of the starting molecule models, build models of the molecules of the ending substances as shown on the yellow side of the mat. Use the *Model Key* to help you make sense of what you are looking at.

Teacher Talk and Actions

Activity 1 **(25 min)**

Whole class:

Emphasize that *Chemical Reaction Mats* represent only the starting and ending substances of the chemical reaction. Students may wonder, for example, why oxygen is a starting substance in rusting, but vinegar (which is also present) is not. And yet, the reverse is true for baking soda—vinegar is a starting substance but oxygen (which is also present) is not.

Partners:

Students complete Step 1.

Introduce the color convention used in the *Chemical Reaction Mats*: Green side shows starting substances and yellow side shows ending substances.

• Once models of starting substances (green side) are built, students should slow down and take time to compare the models they built to what they see on the yellow side of the mat, before taking them apart to build ending substances (yellow side).

Monitor partners to ensure that

• all students are engaged in building models;

• models are built accurately;

• once the starting substance models are built, all other LEGO bricks are put away;

• students take time to think about each chemical reaction after they represent it with models.

2. Answer these questions based on your observations:

   a. What is represented by the *Chemical Reaction Mat?* Use the information on the *Chemical Reaction Mat* to write a word equation for the chemical reaction using the following structure:

   Starting substances ⟶ **Ending substances**

   iron + oxygen gas ⟶ iron oxide

   b. When you built the models of the ending molecules from models of the starting molecules, did you have any leftover LEGO bricks (atoms)?

   No

   c. Did you need any additional LEGO bricks (atoms)?

   No

3. Disassemble your LEGO models.

4. Repeat Step 1, using the *Baking Soda and Vinegar Chemical Reaction Mat.*

5. Answer these questions based on your observations:

   a. What is represented by the *Chemical Reaction Mat?* Use the information on the *Chemical Reaction Mat* to write a word equation for the chemical reaction using the following structure:

   Starting substances ⟶ **Ending substances**

   sodium bicarbonate + acetic acid ⟶ carbon dioxide + water + sodium acetate

   b. When you built the models of the ending molecules from models of the starting molecules, did you have any leftover LEGO bricks (atoms)?

   No

   c. Did you need any additional LEGO bricks (atoms)?

   No

SE p. 47

Teacher Talk and Actions
Activity 1 (cont)
Partners: Students complete Step 4 and answer the questions in Steps 5–7.

d. How would you describe what you did to the LEGO bricks (atoms) in order to make the models of the ending molecules from the models of the starting molecules?

We disconnected the LEGO bricks of the models of molecules of the starting substances and then connected them to different bricks or in different ways to form the models of molecules of the ending substances models.

e. Did you always have to disconnect every LEGO brick (atom) of the models of the starting molecules to make the models of ending molecules? Give a specific example from the two reactions.

No, sometimes groups of bricks stayed together. Like $C_2H_3O_2$ in acetic acid and sodium acetate.

6. Based on what you learned with the models, write a general rule or statement that explains what happens to the atoms in a chemical reaction.

In chemical reactions, atoms rearrange to form molecules of new substances. This means that at least some atoms disconnect and reconnect in new ways.

7. Apply your thinking to predict which ending substances could form when methane gas reacts with oxygen gas from the air. Table 1.16 shows space-filling models for these two starting substances. For each substance shown in Table 1.17, indicate whether it is or is not a possible ending substance and explain why.

Table 1.16. Starting Substances

Starting Substances	
Methane gas CH_4	Oxygen gas O_2
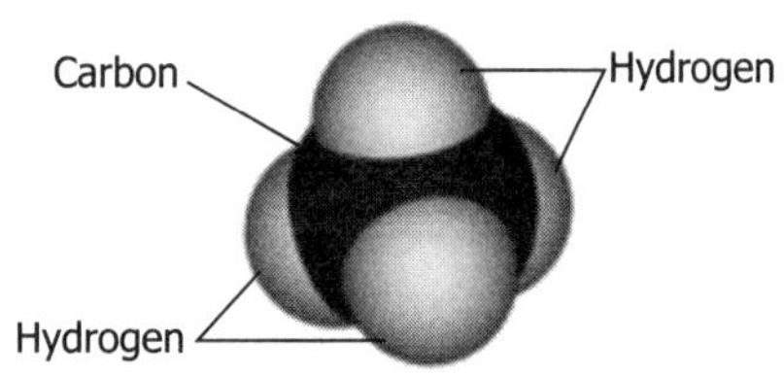	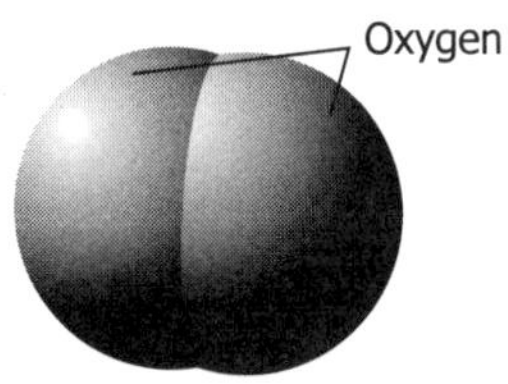

Teacher Talk and Actions
Activity 1 (cont)
Whole class: Lead discussion. Invite students to share ideas about Question 5d and Question 5e. Question 5d. How would you describe what you did to the LEGO bricks (atoms) in order to make the models of the ending molecules from the models of the starting molecules? Question 5e. Did you always have to disconnect every LEGO brick (atom) of the models of the starting molecules to make the models of ending molecules?

Table 1.17. Possible Ending Substances

Possible Ending Substances	**Fill in the blank in each sentence with "is" or "is not," and write why you think this.**
Water H_2O 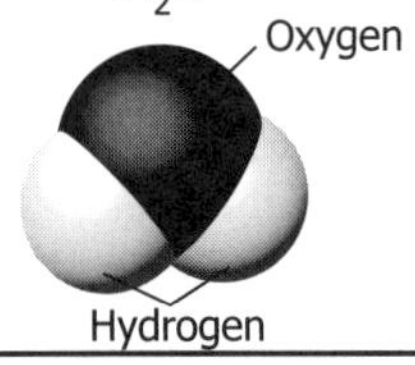	I think water *IS* a possible ending substance because ... *it is made up of hydrogen atoms and an oxygen atom—both of which are in the starting substance molecules.*
Carbon dioxide CO_2 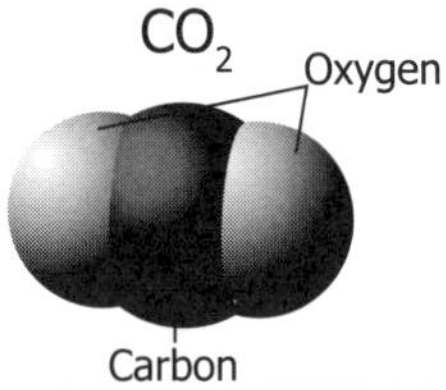	I think carbon dioxide *IS* a possible ending substance because ... *it is made up of a carbon atom and oxygen atoms—both of which are in the starting substance molecules.*
Carbon monoxide CO 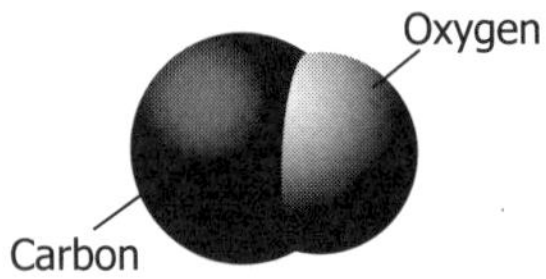	I think carbon monoxide *IS* a possible ending substance because ... *it is made up of a carbon atom and an oxygen atom—both of which are in the starting substance molecules.*
Ammonia NH_3 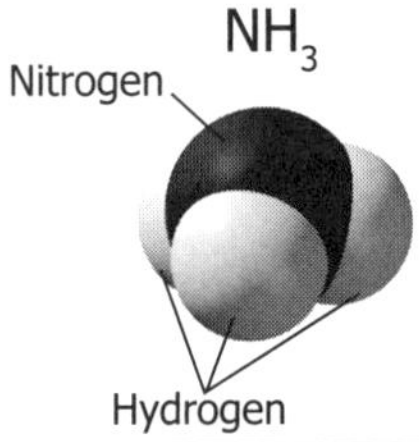	I think ammonia *IS NOT* a possible ending substance because ... *it contains a nitrogen atom, which is not an atom that is found in the starting substance molecules.*

Teacher Talk and Actions
Activity 1 (cont)
Lead discussion. Invite students to share ideas about Step 7. For each ending substance listed in Table 1.17, do the following: • Poll students to see who thinks each substance is/is not a possible ending substance. • Ask for volunteers of each side to explain their thinking.

Science Ideas

The activity in this lesson was intended to help you construct an important idea about chemical reactions. Read the idea below. Notice that the second sentence of Science Idea #5 explains the formation of new substances during chemical reactions in terms of atom rearrangement. In this unit, we will often explain things that we can see, such as the formation of substances like rust or water, in terms of things we can't see—namely, what is happening to the atoms and molecules. We can observe firsthand or infer from data when new substances are produced, but we can't see the atoms rearranging. However, because Science Idea #5 states a general principle that is consistent with a wide range of observations and data, we can use it to reason that whenever we see that one or more new substances are produced, we know that atoms must have rearranged. **From now on, you will be expected to use ideas about atom rearrangement whenever you are asked to explain why one or more new substances form.**

Look back through Lesson 1.5. In the space provided after the science idea, give at least one piece of evidence that supports the idea.

Science Idea #5: During chemical reactions, atoms that make up molecules of the starting substances (called reactants) disconnect from one another and connect in different ways to form the molecules of the ending substances (called products). Because the arrangement of atoms in the products is different from the arrangement of atoms in the reactants, the products of a chemical reaction have different properties from the reactants.

Evidence:

In Lesson 1.5 we represented atom rearrangement in chemical reactions by (1) building models of molecules of the starting substances (reactants) using LEGO bricks to represent atoms, then (2) rearranging those bricks to build models of the molecules of the ending substances (products). The bricks of reactants are arranged differently in the molecules of the products, with no bricks left over and no additional bricks needed. For example, we broke apart models of three oxygen gas molecules (3 O_2) and connected the oxygen atoms to models of four iron atoms (4 Fe) in steel wool to form two molecules of the product iron oxide (2 Fe_2O_3).

When we used LEGO bricks to model the reaction of baking soda and vinegar, we saw that we didn't need to rearrange every atom: The LEGO bricks of $C_2H_3O_2$ from acetic acid ($HC_2H_3O_2$) stayed together when we formed sodium acetate ($NaC_2H_3O_2$).

Teacher Talk and Actions	
Science Ideas	**(5 min)**

Whole class:

Tell students: You have used models to help think about what happens to atoms that make up the reactant molecules during chemical reactions. Now you can compare your ideas with scientists' ideas.

Read the science idea(s) aloud to the class. Add to the list of charted science ideas. Give students time to write down at least one piece of evidence that supports the idea.

Check to be sure that all students understand that what is being modeled is a process (i.e., the rearrangement of atoms to make different molecules).

Remind students that the science ideas will be used in constructing valid explanations.

Activity 2: Constructing Scientific Explanations—Part III

Just as scientists often do, you have used models of chemical reactions to visualize a process that takes place at a scale too small to be seen. The models you have used to help visualize and think about atom rearrangement are based on a great deal of scientific evidence (see Table 1.18). In this activity, you will use what you have learned from the models to help explain what might be happening during chemical reactions.

Table 1.18. Parts of an Explanation and Criteria for Evaluating Its Quality

	What It Is	Explanation Quality Criteria
Question	The question to be answered	
Claim	A statement or conclusion that is intended to respond to the question	• The claim responds to the question.
Science Ideas	Widely accepted scientific ideas, concepts, or principles that can be used to show why the evidence supports the claim	• The science ideas that are listed (or paraphrased) are relevant to the claim. • The science ideas are used to show why the evidence supports the claim. (This involves applying the science ideas to the specific case in the question.)
Evidence	Data that are relevant to and support the claim and can be confirmed by others Data are based on observations about the world that are either made with our senses or measured with instruments.	• The evidence supports the claim. (All the data should be consistent with the science idea, not just the data that are cited as evidence.)
Models	Tools for thinking about the world that can give you ideas about how something might work	• The models are consistent with the science ideas listed. • The models show that the claim is reasonable.

Remember that during Lesson 1.3 we answered the question, Did a new substance form when baking soda and vinegar were mixed, and, if so, where did it come from? Now we have added *models* as another part of a valid explanation. We have also added two new Explanation Quality Criteria to help us think about models. Let's see how one student used models to reason about atoms and molecules in his claim.

Teacher Talk and Actions
Activity 2 **(15 min)**
If the class period is almost over, consider stopping here. If you stop here: **Have students summarize their learning by responding to Pulling It Together Question 1.** Whole class: Read the introductory paragraph aloud to the class. Individuals: Students independently review the parts of the explanation and the Explanation Quality Criteria. Whole class: Discuss the Explanation Quality Criteria with students. For each criterion, you might ask students to refer to the criteria for evaluating explanations developed in Lesson 1.2.

1. Examine the student's notes and how they were put together to construct a valid explanation.

Question	Did a new substance form when baking soda and vinegar were mixed, and if so, where did it come from?
Claim	*Yes, new substances formed that came from a chemical reaction between baking soda and vinegar. I think the Na, H, C, and O atoms making up carbon dioxide, water, and sodium acetate came from the baking soda and vinegar. These atoms originally made up the baking soda and vinegar and must have disconnected and then connected in new ways to form the molecules of the new substances.*
Science Ideas	*Changes during which new substances form are called chemical reactions. The correlation of increasing amounts of ending substances with decreasing amounts of the starting substances provides evidence that the new substances result from an interaction between the starting substances (Science Idea #2).* *New substances form during chemical reactions because atoms that make up molecules of the reactants rearrange to form the molecules of the products, which have different properties from the reactants (Science Idea #5).*
Evidence	*Evidence a new substance was produced: The ending substances have different properties from the starting substances. Baking soda is a white solid, and vinegar is a clear, colorless, smelly liquid. When they were mixed, a colorless, odorless gas was produced that gave a positive limewater test.* *Evidence the carbon dioxide came from the chemical reaction between baking soda and the vinegar: We observed that after baking soda and vinegar were mixed, the smelly liquid and the white powder were gone but a gas that gave a positive limewater test appeared. This test gives evidence that carbon dioxide is present.* *Evidence sodium acetate ($NaC_2H_3O_2$) is also produced: Reactants baking soda ($NaHCO_3$) and vinegar ($HC_2H_3O_2$) have Na, H, C, and O atoms; but carbon dioxide (CO_2) doesn't have any Na or H atoms. The formula on the substance card for sodium acetate ($NaC_2H_3O_2$) shows that it has Na and H atoms.* *Evidence water (H_2O) is also produced: At the end of the chemical reaction, a clear, colorless liquid is present in the plastic bag. However, CO_2 is a gas and $NaC_2H_3O_2$ is a white solid. Water is a clear colorless liquid, and sodium acetate dissolves in water.*

SE p. 52

Teacher Talk and Actions
Activity 2 (cont)
Individuals: Students independently read the parts of the explanation and the student's explanation. Direct students to highlight, underline, or circle in the table and in the student's explanation • where and how models were used to support this explanation, and • which new science idea is used to support this explanation.

Models	Using LEGO models of the molecules of baking soda and vinegar, we showed it was possible to rearrange the LEGO bricks to form molecules of carbon dioxide, water, and sodium acetate. So the Na, H, C, and O atoms making up baking soda and vinegar must have disconnected and connected in new ways to form the molecules of carbon dioxide, sodium acetate, and water.

Explanation: New substances formed from a chemical reaction between baking soda and vinegar. Changes during which starting substances interact to form new substances are called chemical reactions (Science Idea #2). The ending substances can be recognized as new substances because they have different properties from the starting substances (Science Idea #5). Baking soda is a white solid, and vinegar is a clear, colorless, smelly liquid. When they were mixed, a gas appeared that wasn't initially present. The gas gave a positive limewater test. Because a positive limewater test indicates the presence of carbon dioxide, we have evidence that the gas is carbon dioxide. Because carbon dioxide has different properties from vinegar and baking soda, we have evidence of a chemical reaction.

The atoms that make up the new substances (Na, H, C, and O) come from the molecules that made up the baking soda ($NaHCO_3$) and vinegar ($HC_2H_3O_2$). New substances form during chemical reactions because atoms that make up the molecules of the reactants disconnect and reconnect to form the molecules of the products (Science Idea #5). This means that the types of atoms in the products of a chemical reaction are the same as the types of atoms in the reactants. The substance cards of the reactants, which give the chemical formulas of the molecules, show that baking soda ($NaHCO_3$) is made up of Na, H, C, and O atoms, and vinegar ($HC_2H_3O_2$) is made up of H, C, and O atoms. Because both baking soda and vinegar are made up of C and O atoms, the C and O atoms of the CO_2 could have come from either the baking soda or the vinegar. However, since the product (CO_2) doesn't have any Na or H atoms, there must be one or more products whose molecules contain those atoms. The substance card for sodium acetate ($NaC_2H_3O_2$) shows that it has Na and H atoms, and therefore sodium acetate ($NaC_2H_3O_2$) could be another product.

However, we have evidence for yet another product. There is a clear colorless liquid present at the end of the chemical reaction, and the substance cards show that carbon dioxide is a gas and sodium acetate is a white solid. Since neither carbon dioxide nor sodium acetate is a clear, colorless liquid, there must be another product. Water is a clear, colorless liquid, so water could be another product.

We don't see a white solid at the end of the reaction. However, we observed before that sodium acetate dissolves in water. So maybe water is also a product and the clear, colorless liquid is actually sodium acetate dissolved in water. To confirm, we would need to separate the substances and determine their properties.

Teacher Talk and Actions
Activity 2 (cont)
Small groups: Teams evaluate the explanation, using the Explanation Quality Criteria. Whole class: Lead discussion. Invite student volunteers to share their group's explanation with the whole class, or pose these questions: • Does the explanation fail to meet any of the Explanation Quality Criteria? • If yes, what are some suggestions for improving the explanation? Students should come to consensus that this is a valid explanation because it meets the Explanation Quality Criteria.

2. With your team, evaluate the student's explanation. Discuss the Explanation Quality Criteria listed in Table 1.18. Underline the part of the student's explanation where models were used to help in the student's reasoning.

3. Now write an explanation to the question, Did a new substance form when steel wool was exposed to oxygen, and if so, where did the atoms that make up the new substance come from?

Use the table to help you organize your thoughts. Remember to reason using atoms and molecules in your explanation.

Question	Did a new substance form when steel wool was exposed to oxygen, and if so, where did the atoms that make up the new substance come from?
Claim	*Yes, rust is a new substance made of Fe_2O_3 molecules. The O atoms came from O_2 molecules of oxygen gas. The Fe atoms came from Fe atoms from the steel wool.*
Science Ideas	*Changes during which new substances form are called chemical reactions. The correlation of increasing amounts of ending substances with decreasing amounts of the starting substances provides evidence that the new substances result from an interaction between the starting substances (Science Idea #2).* *New substances form during chemical reactions because atoms making up the molecules of the reactants rearrange to form the molecules of the products (Science Idea #5).*
Evidence	*Evidence a new substance was produced: A reddish-brown solid was produced that wasn't there at the beginning.* *Evidence something in the air was used: As the reddish-brown solid formed, the balloon was pushed into the flask.* *Evidence the "something" was oxygen: There was a positive splint test before the chemical reaction and a negative splint test after the chemical reaction had occurred.* *Evidence the steel wool was used: After the steel wool had been exposed to oxygen for an hour, there was still gray metal underneath the rust, but after a week in air, all of the gray metal was gone, and more of the crumbly, reddish-brown solid was present.* *Data on the cards showed the correct formulas for oxygen gas (O_2), iron (Fe), and rust (Fe_2O_3).*

Teacher Talk and Actions
Activity 2 (cont)
(Intentionally left blank)

Models	Using models that correctly represent the molecular composition of each substance, we observed that it is possible to make molecules of rust (Fe_2O_3) from molecules of oxygen (O_2) and iron atoms (Fe) just by disconnecting the bricks (atoms) and connecting them differently.

Explanation:

Yes, I think that the rust is a different substance and that it was produced during a chemical reaction between the reactants iron and oxygen gas. If a substance with different properties is produced then a chemical reaction has occurred (Science Idea #2). My evidence is that rust is a reddish-brown solid and neither iron (a grayish-black solid) or oxygen (a colorless, odorless gas that gives a positive splint test) is a reddish-brown solid. Since rust has different properties from iron and oxygen, it is a different substance. New substances form during chemical reactions because atoms making up the molecules of the reactants rearrange to form the molecules of the products (Science Idea #5). Our models showed us that it is possible to make the molecules of Fe_2O_3 from the atoms making up Fe and O_2 by disconnecting the oxygen atoms in O_2 and connecting them to Fe atoms to make Fe_2O_3 molecules. What we observed with our models is consistent with Science Idea #5: Fe_2O_3 molecules form during chemical reactions because atoms making up the Fe and O_2 rearrange. So I think rust formed by rearranging atoms from iron and oxygen.

4. Discuss how your explanation meets the Explanation Quality Criteria.

I answered the question: Yes, the rust is a different substance.

I cited science ideas, and used them to show how this example was an instance of the science idea.

I provided evidence about the properties of rust and iron and oxygen.

I talk about how models could be used to show that the atom rearrangement was possible.

Teacher Talk and Actions
Activity 2 (cont)
Explanation Quality Criteria met: (During discussion, you can ask students to explain their conclusions.) • *The answer responds to the question: The question asks if a new substance formed and the answer says that a new substance, rust, formed.* • *Science Ideas #2 and #5 are relevant because that is what happened with iron rusting: A substance was produced with different properties and molecules of product can be made by rearranging atoms of reactant molecules.* • *The evidence supports the answer: The evidence has to establish that a substance with different properties is produced. The statement that a reddish-brown solid was produced that wasn't there before provides evidence that a substance with different properties was produced.* • *Our models are consistent with Science Idea #5: Our models showed us that it is possible to make the molecules of Fe_2O_3 by disconnecting the oxygen atoms in O_2 and connecting them to Fe atoms.* • *Science Ideas #2 and #5 are used to support the answer.*

Pulling It Together

Work on your own to answer these questions. Be prepared for a class discussion.

1. Summarize your learning in this lesson by answering the Key Question, **How are the molecules of the ending substances made?** Using atoms and molecules, explain your answer in a way that you would to a friend who missed class today.

The molecules of the ending substances are made from the atoms of the starting substances. Atoms that make up the molecules of starting substances (reactants) disconnect and then connect to different atoms to form the molecules of ending substances (products).

2. In this unit, you have also seen the chemical reaction between hexamethylenediamine and adipic acid. The chemical formula for hexamethylenediamine is $C_6H_{16}N_2$, and the chemical formula for adipic acid is $C_6H_{10}O_4$. Based on what you now know about atom rearrangement in chemical reactions, what atoms do you predict will be present in the products (ending substances)? Support your answer.

The product molecules will be made up of C, H, O, and N atoms. The molecules of the products are made from the atoms of the reactants, so the product molecules will be made up of the same atoms as the reactant molecules. During chemical reactions, atoms that make up the molecules of the reactants rearrange to form the molecules of the products (Science Idea #5).

Teacher Talk and Actions	
Pulling It Together	**(25 min)**

Individuals:

Students answer Pulling It Together questions on their own, in detail.

Whole class:

Lead a class discussion. Invite students to share ideas about Pulling It Together Question 1 (the Key Question).

Make sure students are being clear about the relationship between molecules and atoms (atoms make up molecules).

Direct students to look back at their initial ideas about the Key Question and consider how their ideas have changed.

3. Hydrogen peroxide is a clear, colorless liquid. When you apply it to a cut or scrape on your skin, you will see bubbles and you may hear a “fizzing” sound. Figure 1.3 uses LEGO models to represent hydrogen peroxide on the left. On the right are models showing what happens to hydrogen peroxide when you apply it to a cut or scrape on your skin. Your teacher will project a color version of Figure 1.3 so that you can see the models more clearly.

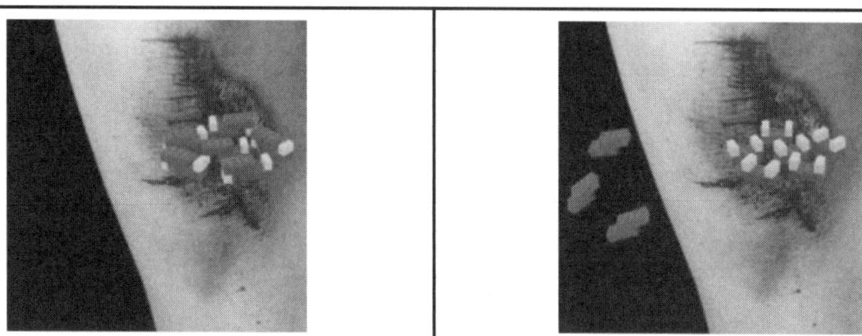

Before the change	After the change

Figure 1.3. Hydrogen Peroxide Before (left) and After (right) Change

Do you think a chemical reaction occurs when hydrogen peroxide is put on a wound? Explain why you do or do not think so in terms of atoms and molecules. Your explanation should use science ideas, evidence, and models to support your claim. (Use the table for your notes.)

Question	Is the change hydrogen peroxide undergoes an example of a chemical reaction?
Claim	*Yes, a chemical reaction occurs when hydrogen peroxide is put on a wound.*
Science Ideas	*Option 1: During chemical reactions, atoms that make up reactant molecules disconnect and connect in different ways to form the product molecules (Science Idea #5).* *Option 2: Because no two substances are made up of the same arrangement of atoms, no two substances have the same set of properties (Science Idea #4), AND the ending substances of a chemical reaction can be recognized as new substances because they have different properties from the starting substances (Science Idea #2).*
Evidence	*Option 1: Scientific models that accurately represent the molecules show that after the change, two new molecules are present—O_2 and H_2O. Atoms are arranged differently in the products compared with the reactant molecules.* *Option 2: Scientific models that accurately represent the molecules show that the formulas of the ending substances (O_2 and H_2O) are different from the formulas of the starting substance (H_2O_2).*

Teacher Talk and Actions

Pulling It Together (cont)

Whole class:

Pulling It Together Question 3 can be used as an embedded assessment.

- Poll students to see who thinks the change is a chemical reaction or not a chemical reaction.
- Ask for volunteers of each "side" to explain their thinking. As they do, ask:
    - Is the science idea you listed relevant to your answer? Why?
    - Is the evidence consistent with the science idea? Why?
    - Do the models show the answer is reasonable and consistent with the science idea? Why?

Note that there are two different ways that students can approach the question:

- Option 1 involves reasoning from Science Idea #5 and evidence from different formulas that a chemical reaction has occurred because atoms rearranged. This option is the most straightforward and gives students the opportunity to use the idea of atom rearrangement.
- Option 2 involves reasoning from Science Idea #4 and Science Idea #2 and evidence from different formulas that a chemical rcaction has occurred because a substance with different properties was produced. While perfectly acceptable, monitor students who use this option to be sure that they are using Science Idea #5 to make sense of nylon formation in Lesson 1.6 and in explaining the Statue of Liberty question in Lesson 2.3.

Make sure students are being clear about the relationship between molecules and atoms (atoms make up molecules), what changes during chemical reactions (molecules), and what stays the same in chemical reactions (number and types of atoms).

Models	*Option 1: It is possible to form LEGO models of 2 H_2O and O_2 from 2 H_2O_2 by disconnecting 1 O brick from 2 H_2O_2 (leaving 2 H_2O) and then connecting the 2 O bricks to form O_2.* *Option 2: Because the LEGO models of the ending substances (H_2O + O_2) are different from the LEGO models of the starting substance (H_2O_2), the ending substances have different formulas from the starting substance.*

Explanation:

Option 1: Yes, I think that a chemical reaction occurs when hydrogen peroxide is put on a wound. During chemical reactions, atoms of reactant molecules rearrange to form product molecules (Science Idea #5). Scientific models (based on data from experiments) show that two new molecules are produced—H_2O and O_2. Our models show that it is possible to form H_2O and O_2 by disconnecting an O brick from 2 different H_2O_2 models (which would leave 2 H_2O models) and then connecting the 2 O bricks to form O_2. So I think a chemical reaction occurred to form H_2O and O_2 from H_2O_2.

Option 2: Yes, I think a chemical reaction occurs when hydrogen peroxide is put on a wound. Because no two substances are made up of the same arrangement of atoms, no two substances have the same set of properties (Science Idea #4). The LEGO models show that H_2O and O_2 have a different arrangement of atoms from H_2O_2. Therefore H_2O and O_2 must have different properties from H_2O_2.

The ending substances of a chemical reaction can be recognized as new substances because they have different properties from the starting substances (from Science Idea #2). Since the ending substances (H_2O and O_2) must have different properties from the starting substance (H_2O_2), they must be different substances. Therefore, a chemical reaction occurred.

a. Underline the part of your explanation where you used models in your reasoning.

b. How does your explanation meet the Explanation Quality Criteria? Give details. (See Table 1.18 on p. 51 for a reminder of what the criteria are.)

I answered the question: Yes, a chemical reaction occurs.

I cited science ideas, and used them to show how this example was an instance of the science idea.

I provided evidence about the properties of rust and iron and oxygen.

I talked about how models could be used to show that the atom rearrangement was possible.

Teacher Talk and Actions
Pulling It Together (cont)
The following are points that should be discussed with the whole class: Option 1: • The answer responds to the question: The question asks if a chemical reaction occurred, and the answer says that chemical reaction occurred. • The science idea(s) are relevant: Science Idea #5 is relevant because it says you can use atom arrangement to decide whether or not a chemical reaction has occurred. • The evidence supports the answer: The evidence has to establish that atoms of the reactant molecules are connected differently from atoms of the product molecules. H_2O and O_2 are made of the same atoms as H_2O_2, but they are connected differently. This provides evidence that atom rearrangement occurred. • Our models are consistent with the idea that new substances form during chemical reactions because atoms of the reactants rearrange to form products: We could take apart two H_2O_2 molecules and use the H and O atoms to make two H_2O and one O_2 molecules. • The idea that atom rearrangement occurred is used to show why the evidence supports the answer: It restates the science idea in terms of the specific example of the production of H_2O and O_2 molecules with different arrangements of the atoms of H and O atoms from H_2O_2. Option 2: The answer responds to the question: Same as above The science idea(s) are relevant: Science Idea #2 is relevant because it says you can use properties to decide whether or not a chemical reaction has occurred. Science Idea #4 is relevant because it links properties to atom arrangement. The evidence supports the answer: The evidence has to establish that a different substance was produced. The LEGO models show that the ending substances have different formulas (H_2O and O_2) from the starting substance (H_2O_2). Since their formulas are different, their properties must be different. (Note, students could find properties of hydrogen peroxide on the internet and use that data as evidence that the ending substances have different properties from the starting substances.) And since the properties of water and oxygen are different from the properties of hydrogen peroxide, a new substance was produced and a chemical reaction occurred. Review students' explanations after class and discuss the next day. Be sure to discuss an example response that uses Option 1. You don't need to discuss Option 2 unless some of your students take this approach.

4. As a pot of water is heated on a stove, bubbles of gas form in the liquid. Figure 1.4 uses LEGO models to represent the molecules of water in the pot on the left. On the right are models of the molecules inside the bubbles that form.

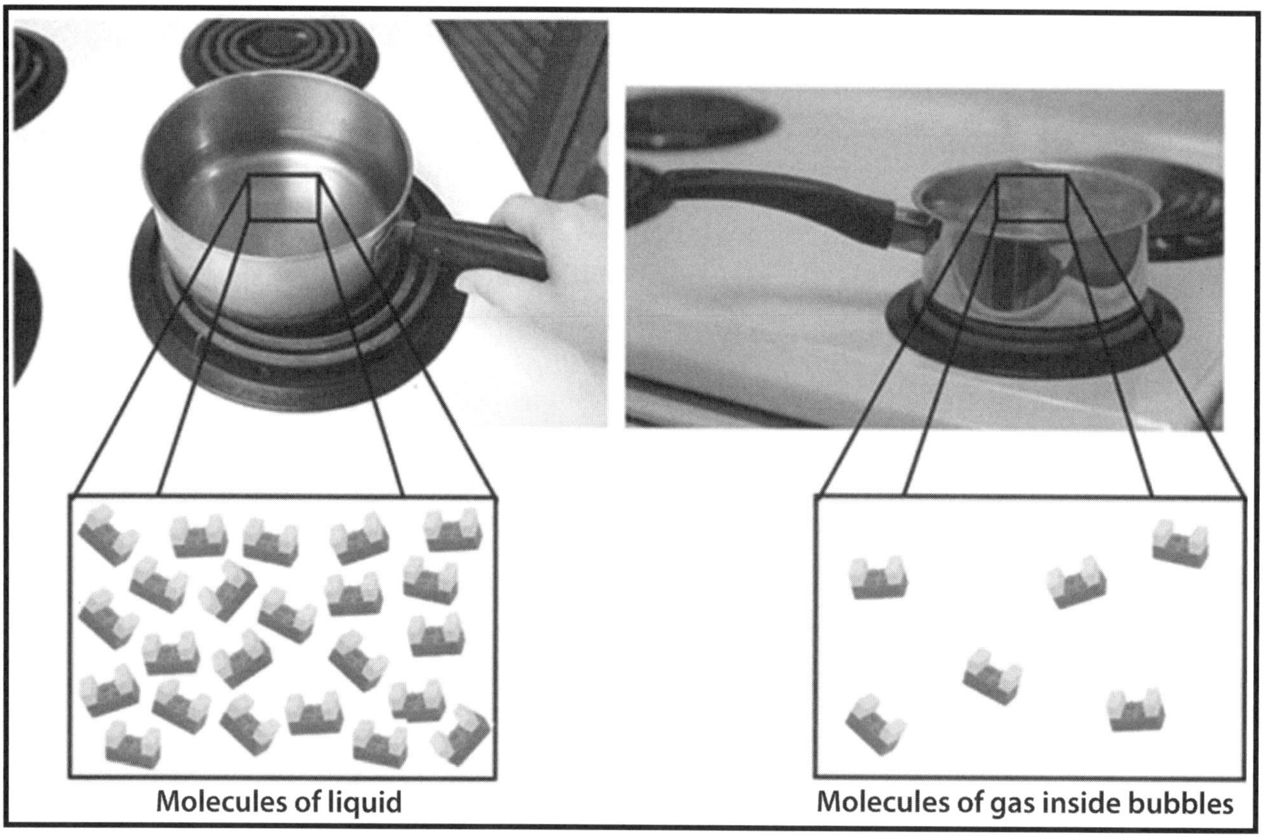

Figure 1.4. Water Being Heated on Stove

Do you think a chemical reaction occurs when water is heated on a stove? Explain why you do or do not think so, using science ideas, evidence, and models to support your claim. (Use the table for your notes.)

Question	Does a chemical reaction occur when water is heated on a stove?
Claim	*No, a chemical reaction does not occur.*
Science Ideas	*During chemical reactions, atoms of reactant molecules rearrange to form product molecules OR If no atoms rearrange then no chemical reaction occurs (Science Idea #5).*

Teacher Talk and Actions

Pulling It Together (cont)

Whole class:

Pulling It Together Question 4 can be used as an embedded assessment.

- Poll students to see who thinks the change is a chemical reaction or not a chemical reaction.

- Ask for volunteers of each "side" to explain their thinking. As they do, ask:
    - Is the science idea you listed relevant to your answer? Why?
    - Is the evidence consistent with the science idea? Why?
    - Do the models show the answer is reasonable and consistent with the science idea? Why?

Make sure students are being clear about the relationship between molecules and atoms (atoms make up molecules), what changes during chemical reactions (molecules), and what stays the same in chemical reactions (number and type of atom).

Review students' explanations after class.

Evidence	*Scientific models show that the arrangement of atoms of the ending substance (HOH) is the same as the arrangement of atoms of the starting substance (HOH), even though HOH molecules are farther apart from one another.*
Models	*LEGO models of starting and ending substances show that atom rearrangement doesn't happen.*

Explanation:

I do not think that a chemical reaction occurred when the water was heated. Atoms making up the molecules of reactants are rearranged to form molecules of products during a chemical reaction (Science Idea #5), so to conclude that a chemical reaction occurred we need evidence that new molecules formed. Scientific models show that the arrangement of atoms is the same (HOH) in both the liquid and gas. The models show that atoms of the starting substance don't have to rearrange to form the molecules of the ending substance. If atoms did not rearrange when the liquid was heated, then a chemical reaction did not occur.

Teacher Talk and Actions
Closure and Link **(4 min)**
Whole class: Summarize the lesson: We used models to help us visualize what happens to atoms during chemical reactions. We learned that during chemical reactions, atoms making up molecules of the starting substances rearrange and connect in different ways to make molecules of new substances. We learned that we didn't always have to disconnect every LEGO brick of the starting substance models to make the ending substance models. Link to the next lesson: We will apply both of these ideas to learning about making very large molecules from small molecules in the next lesson. In Chapters 3 and 4, we will use this idea to help you explain how living things grow and repair.

Lesson Guide
Chapter 1, Lesson 1.6
Representing Chemical Reactions That Produce Large Molecules

Focus: The lesson uses ball-and-stick models to represent the formation of nylon (polymer) during a chemical reaction in which only a few atoms of hexamethylenediamine and adipic acid (monomers) rearrange. The lesson confronts the common ideas that during chemical reactions that make very large molecules, starting substances are merely "stuck together" (i.e., no atom rearrangement occurs) or they are completely broken down into a pile of individual atoms (i.e., every atom disconnects) that then connect in different ways. Understanding polymerization reactions is essential to understanding biological growth.

Key Question: How can small molecules be used to make very large molecules?

Target Idea(s) Addressed

- Atoms form molecules that range in size from two to thousands of atoms.
- Very large molecules called polymers can be formed by reacting small molecules (monomers) together.
- Because monomers can react in two places, it is possible for each monomer to react with two other monomers to form long polymer chains.
- Each time a monomer is added to the chain, atoms are rearranged and each time a monomer is added to the chain, a water molecule is formed.

Science Practice(s) Addressed

- Use a model to describe phenomena.
- Apply scientific ideas, principles, or evidence to construct an explanation for real-world phenomena.

Materials

Activity 1: Copymaster: *Chemical Reaction Mat:* Hexamethylenediamine and Adipic Acid, HGS ball-and-stick model kit, wet or dry erase markers, trays for working with models on tables, colored pens or pencils

Advance Preparation

Organize materials in advance so that each set of partners or table teams has materials when class begins.

Phenomena, Data, or Models	Intended Observations	Purpose	Rationale or Notes
Activity 1 Ball-and-stick models to represent the chemical reaction that links hexamethylenediamine and adipic acid monomers to form nylon polymer and water molecules	• Balls (atoms) of the models of hexamethylenediamine and adipic acid can be disconnected and connected in different combinations to make models of nylon and water, with no balls left over and no additional balls needed. • When balls are rearranged, monomers are linked together and a molecule of water is made. • Only a few balls of the monomers (O and H atoms from groups on either end of the molecule) rearrange to make models of the new substances. • The balls can be rearranged at two places on each monomer, so monomers can be linked together in long chains.	Students visualize the rearrangement of atoms (and forecast the conservation of atoms) of the starting substances to account for why the new substances that form during chemical reactions have different properties.	The modeling activity challenges common incorrect ideas that during chemical reactions that make very large molecules, starting substances are merely "stuck together" or they are completely broken down into a pile of atoms that self-organize into large molecules. Rather, chemical reactions are involved, during which atoms are rearranged in very specific ways.
Pulling It Together Ball-and-stick models representing segments of two biological polymers: • Cotton (cellulose) • Spider silk (protein) Ball-and-stick models representing monomers that react to form Kevlar polymer	• Cellulose (making up cotton) is a polymer made up of a single, repeating, monomer consisting of C, H, and O atoms with some C atoms and an O atom linked together in a ring. • Spider silk is a polymer made up of different monomers consisting of C, H, O, and N atoms. The monomers appear to have some configurations of atoms in common (e.g., a repeating O-C-N-) and parts that make them different from one another.	Students use accurate models of polymers to identify possible monomers that make them up.	The contexts challenge the common idea that all substances, in this case polymers, are made up of the same type of molecule. Rather, the term *polymer* is the generic term for large molecules that are formed through specific reactions that link monomers in long chains. Students will later learn that protein polymers are made up of specific kinds of monomers called amino acids, and plant cellulose polymers are made up of glucose monomers. Kevlar formation is provided as an example of polymer formation that produces HCl rather than H_2O.

Lesson 1.6—Representing Chemical Reactions That Produce Large Molecules

What do we know and what are we trying to find out?

You have used LEGO bricks to represent (or model) what happens when atoms from small molecules rearrange to form other small molecules. But some kinds of molecules are very large (made up of thousands of atoms!). Many human-made materials (for example, nylon, rubber, and plastics) that we see and use each day are made up of very large molecules.

You may not know this, but your body structures, such as muscles and skin, are made up mostly of very large molecules. Plant body structures, such as stems and roots, are also made up of very large molecules. Understanding how very large molecules form from small molecules will eventually help us explain how a puppy and sequoia tree grow, how a wound heals, and how a lizard regrows its tail.

Use what you know about chemical reactions to answer the Key Question.

Key Question: How can small molecules be used to make very large molecules?

If the large molecules have different properties from the small molecules, then they must be made during chemical reactions. That means that atoms making up the small molecules must disconnect and connect in new ways to make the large molecules. Maybe all the atoms of the small molecules come apart and reconnect in different ways to form the large molecules, or maybe what happens is like what we saw in the reaction between baking soda and vinegar where a group of atoms ($C_2H_3O_2$) stayed together, so only a few atoms actually rearranged. I'm not really sure why sometimes small molecules would react to make other small molecules and other times small molecules react to form large molecules. Maybe there's something different about the small molecules that we haven't learned yet.

Teacher Talk and Actions
What do we know and what are we trying to find out? (3 min)
Whole class: Read text under "What do we know and what are we trying to find out?" aloud to the class.
Key Question (5 min)
Individuals: Students respond to the Key Question: **How can small molecules be used to make very large molecules?** Whole class: Lead discussion. Invite students to share ideas about the Key Question. The goal is primarily eliciting and probing student ideas here. The response listed contains statements that would be reasonable for students to make at this point. When students share their answers, you can challenge answers that are not consistent with what students have learned in previous lessons. However, do not indicate if their answers are right or wrong. Make it clear that students are expressing ideas based on what they now know, but those ideas may change based on what they will see in the lessons. (Remind students of this regularly.)

Activity 1: Representing Nylon Formation With Models

Materials
For each team of students
Chemical Reaction Mat: Hexamethylenediamine and Adipic Acid
HGS ball-and-stick model kit
Wet or dry erase markers

In this lesson, you will model the chemical reaction that produces nylon. Pay careful attention to what and how new substances form during the reaction. You will use ball-and-stick models that are more like the ones that scientists use than the models you have built with LEGO bricks in other lessons. These ball-and-stick models represent atoms using balls of the same color as the LEGO bricks, but they use sticks to connect atoms. Ball-and-stick models make it easier to represent large molecules because the ball-and-stick models do not fall apart as easily as LEGO models. We will use ball-and-stick models any time we want to represent molecules that are made up of chains of carbon atoms, such as hexamethylenediamine and adipic acid.

Procedures and Questions

1. Working with a partner, use the *Hexamethylenediamine and Adipic Acid Chemical Reaction Mat* and HGS ball-and-stick model kit to complete the steps below.
   a. Use ball-and-stick model pieces to build models of the molecules of the starting substances shown on the green side of the mat. For each molecule, do the following:
      i. Attach the six carbon atoms in a chain first.
      ii. Add the oxygen or nitrogen atoms to the carbon atoms at either end of the carbon atom chain.
      iii. Then add hydrogen atoms to the carbon, oxygen, and nitrogen atoms as shown on the mat. (The orientations of atoms—pointing toward you or away from you—are not important.)
   b. Once you have built the models of the starting molecules exactly as they are shown on the mat, put all unused ball-and-stick model pieces back in the model kit and close the kit.
   c. Turn the mat over to the yellow side, which shows you how to build your models of the ending molecules. Before you begin building, look carefully at the ending molecules shown on the mat so that you can accomplish the following tasks:
      i. Identify anywhere you are going to break a connection. Use a wet or dry erase marker to draw an *X* on the mat over any connections you are going to break.

         Lay a pencil or pen across these same connections on the ball-and-stick models you built to show which ones will break.

Teacher Talk and Actions	
Activity 1	**(30 min)**

Whole class:

Read introductory text.

Introduce students to the ball-and-stick model kits. Remind students that these models use the same color scheme as the LEGO model but use sticks to connect the balls.

Partners:

Students complete Step 1a together. Each student should make a model of either hexamethylenediamine or adipic acid such that a pair of students makes one of each small molecule.

Monitor partners to ensure that

- all students are engaged in building models;
- models are built accurately;
- once the starting substance models are built, all other model pieces are put away; and
- students take time to think about each chemical reaction after they represent it with models.

ii. Identify where you are going to form any new connections. Use the marker to circle atoms and draw lines on the mat showing any new connections you are going to make. Identify the atoms in the ball-and-stick models that you built to show how atoms will connect.

iii. Have your teacher check your mat and give the OK to make these changes to the models.

d. Use only the ball-and-stick model pieces of the starting molecule models to build models of the ending molecules as shown on the yellow side of the mat.

2. Answer these questions based on your observations:

a. What is represented on the *Chemical Reaction Mat?* Use the information on the *Chemical Reaction Mat* to write a word equation for the chemical reaction.

Starting substances ⟶ **Ending substances**

adipic acid + hexamethylenediamine ⟶ nylon + water

b. When you built the models of the ending molecules from models of the starting molecules, did you have any leftover balls (atoms)?

No

c. Did you need any additional atoms?

No

d. Did you have to disconnect every atom of the models of the starting molecules to make the models of the ending molecules?

No

e. During this chemical reaction, when you connected the two different small molecules together, what other molecule did you form?

We also formed a water molecule, H_2O.

Teacher Talk and Actions
Activity 1 (cont)

Small group/partners:

Students follow prompts in Step 1c to mark up the *Chemical Reaction Mat* and translate those markings to manipulations they will make with the physical models. OK their mats before they begin manipulating the models.

Students discuss and respond to questions in Step 2.

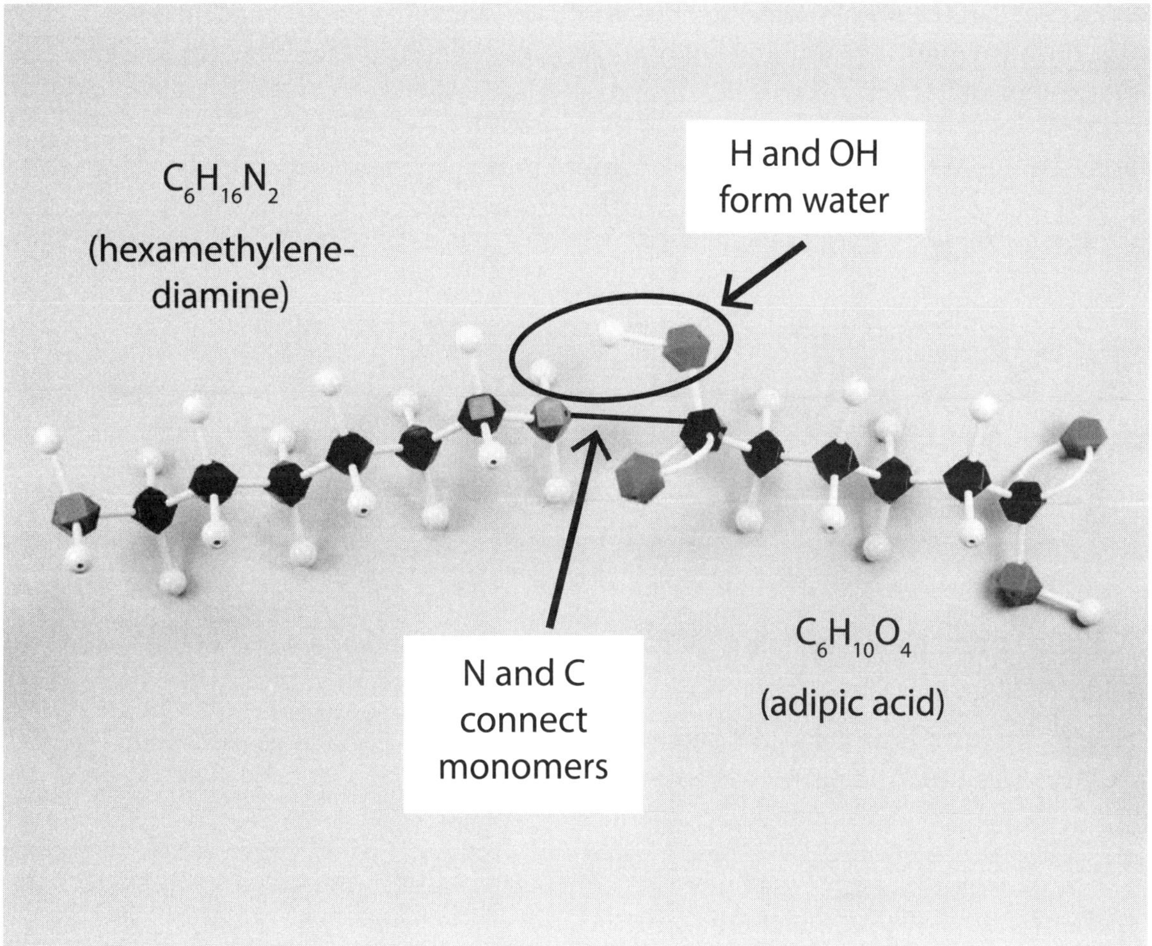

f. Where did the H and OH atoms come from that linked together to form the water molecule? Be as specific as you can.

The O and H came from adipic acid and the H came from hexamethylenediamine.

g. What atoms from each small molecule link together? Be as specific as you can.

The C atom on the end of adipic acid and the N atom on the end of hexamethylenediamine are used to link the two monomers together. The C atom is the same one that the O and H come off of to form water. The N atom is the same one that the H comes off of to form water.

You just modeled the basic chemical reaction that happens between two small molecules. But the small molecules can react in two places, not just on one end as you modeled. So if you had many more of the small molecules, you could link them end to end to form a large molecule. We call the large molecules *polymers* (*poly* means many and *mer* means part) because they are made up of many small molecules called *monomers* (*mono* means one and *mer* means part).

The image in Figure 1.5 shows a longer segment of nylon polymer in which each of the monomers at the end has reacted with another monomer. Color the O atoms red and the N atoms blue.

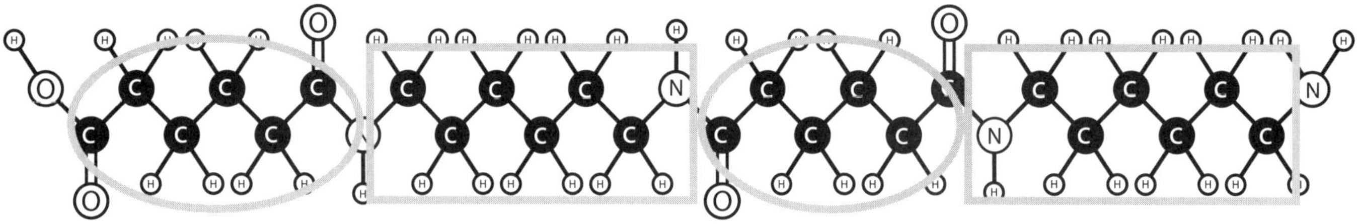

Figure 1.5. Segment of a Nylon Polymer

3. Examine the segment of nylon polymer in Figure 1.5 and answer these questions with your team.

a. How many adipic acid monomers were used to make this polymer chain? Draw circles around the parts that come from adipic acid.

Two

b. How many hexamethylenediamine monomers were used to make this polymer chain? Draw rectangles around the parts that come from hexamethylenediamine.

Two

c. How many water molecules were made?

Three

Teacher Talk and Actions
Activity 1 (cont)
Lead discussion. Invite students to share ideas about the questions in Steps 2f and 2g. Invite students to show the chemical reaction using models. Select two students to conduct a talk-aloud as they demonstrate the chemical reaction with models. Whole class: Read the paragraph aloud to the class. Small group/partners: Students complete Step 3. <u>Science Note to Teacher:</u> The COOH and NH_2 groups in the monomers used to form nylon are analogous to the groups that participate in atom rearrangement when amino acids form proteins (Chapter 4). Helping students identify these groups now will help them in their work later.

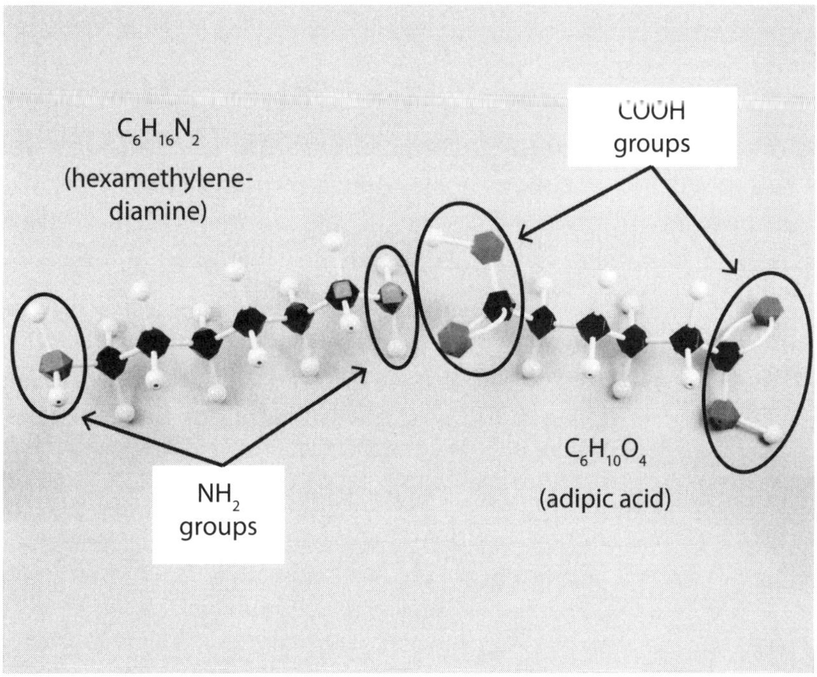

4. Imagine that we connected all of the class's monomers together one at a time to form a single polymer chain with alternating monomers. Answer these questions about the imaginary polymer:

   a. How many total monomer models did the class make?

   Number will vary for each class

   b. How many of each type of monomer were made?

   Number will vary for each class (X adipic acid, X hexamethylenediamine)

   c. How many water models would be made by the class if all of the monomer models were linked together in a chain?

   Number will vary; will be (X + X − 1) water

   d. Write a word equation for the formation of this chain.

Starting substances	→	**Ending substances**
X adipic acid + X hexamethylenediamine	→	*nylon + (X + X − 1) water*

5. Take apart and put away your models.

If a 1,000-mer were made of real atoms instead of ball-and-stick models, it would still be too small for you to see with even the highest-powered microscope. Polymers are extremely large molecules compared with other molecules that you have studied, like water. However, it still takes extremely large numbers of polymer molecules to make up visible amounts of these substances. For example, nylon thread is actually many long polymers twisted together. Just 1 millimeter of nylon thread contains about 100,000,000,000,000,000 monomer molecules. If that many real monomers were represented with ball-and-stick models, the models would fill up more than 50 billion (50,000,000,000) classrooms!

Teacher Talk and Actions
Activity 1 (cont)
If the class period is almost over, consider stopping here. If you stop here: **Have students summarize their learning by responding to Pulling It Together Question 1.** **Assign Step 4 and Pulling It Together Question 2 for homework, with discussion to follow in class tomorrow.** Whole class: Students work together to complete Step 4 based on class data gathered during the activity. Science Note to Teacher: Answers will vary by class, but generally, the number of water molecules that are produced from the chemical reaction among monomers should be one less than the number of total monomers.

Science Ideas

Lesson 1.6 was intended to expand your understanding of atom rearrangement to include polymer formation. Read the idea below. Remember that because science ideas are consistent with a wide range of relevant evidence, you are justified in applying them to new observations and data. Look back through Lesson 1.6. In the space provided after the science idea, give at least one piece of evidence that supports the science idea.

Science Idea #6: Very large molecules called polymers can be formed by reacting small molecules (monomers) together. Because monomers can react in two places, it is possible for each monomer to react with two other monomers to form long polymer chains. Each time a monomer is added to the chain, atoms are rearranged and another molecule, typically water, is formed.

Evidence:

In Lesson 1.6 we used models to represent nylon polymers forming from linking adipic acid and hexamethylenediamine monomers. This happens when groups of atoms (made up of oxygen and hydrogen or nitrogen and hydrogen) are rearranged at two places on each monomer. Water forms when the oxygen atom and hydrogen atom from adipic acid link with a hydrogen atom (from the nitrogen) of hexamethylenediamine. The carbon from adipic acid links to the nitrogen atom of hexamethylenediamine. Because this happens at two places on each monomer molecule, large numbers of monomers can be linked end to end to form very large polymer molecules.

Teacher Talk and Actions

Science Ideas **(5 min)**

Whole class:

Tell students: You have used models to help develop new ideas about how very large molecules (polymers) are made from small molecules (monomers). When we read these science ideas, look closely for details that we may not have noticed before.

Read the science idea(s) aloud to the class. Ask students to share what they wrote to expand the idea to include polymer formation. Add a sentence to Science Idea #6.

Emphasize that

- monomers often contain carbon atoms arranged in chains or rings, and
- groups of atoms on two ends of the monomer can participate in atom rearrangement. Often the groups have COOH and NH_2 atoms.

Pulling It Together

Work on your own to answer the following questions. Be prepared for a class discussion.

1. Answer the Key Question: **How can small molecules be used to make very large molecules?**

   If each small molecule has two groups of atoms that can react (one on each end), then each small molecule (monomer) can react with two other monomers and many monomers can react to form large molecules (polymers).

2. Figures 1.6 and 1.7 show segments of polymer molecules of cotton and spider silk. Color the O atoms red and the N atoms blue. Look for repeating patterns in each polymer molecule. Draw brackets around the monomers for each polymer molecule.

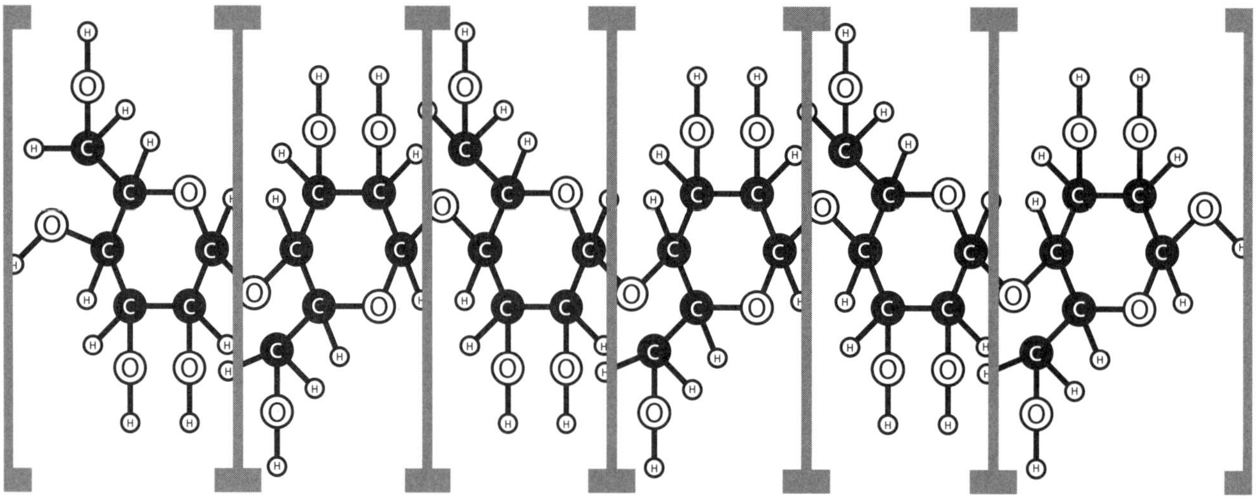

Figure 1.6. Segment of a Cotton Polymer Molecule

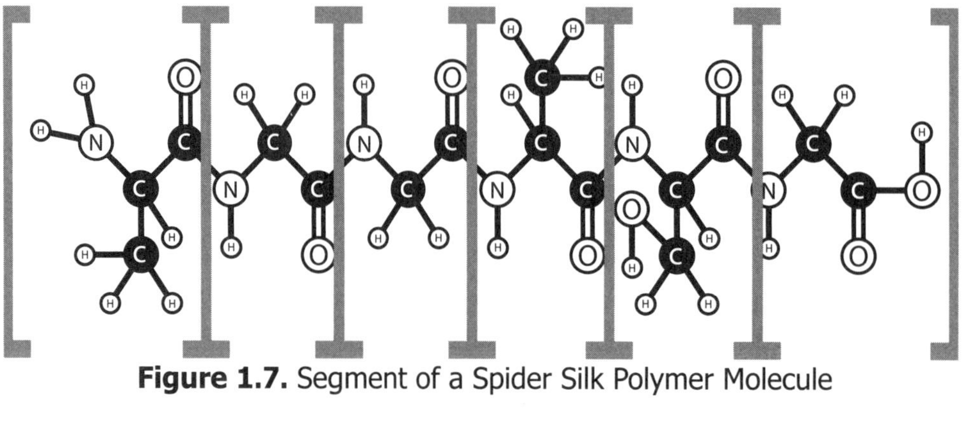

Figure 1.7. Segment of a Spider Silk Polymer Molecule

Teacher Talk and Actions

Pulling It Together **(12 min)**

Individuals:

Students answer Pulling It Together Question 1 on their own.

Direct students to look back at their initial ideas about the Key Question and consider how their ideas have changed.

Students answer Pulling It Together Questions 2 and 3 on their own and then discuss with their groups.

Whole class:

Lead a class discussion. Invite students to share ideas about possible monomers.

Science Note to Teacher:

Plant cellulose is made up of one monomer, called glucose. Because glucose molecules alternate orientations, they may appear as two different monomers to students.

Spider silk protein is made up of several different monomers, called amino acids. To identify where one amino ends and another starts, begin at the left "NH_2" group and look for the repeating "backbone": [N(H)-CH(R)-C(=O)] where R could be H or another carbon chain, and (=O) is an oxygen double-bonded to the C atom in the backbone.

It is OK if students have difficulty identifying the monomers. In Chapters 3 and 4, students will have more practice working with amino acid and glucose monomers.

3. Kevlar is a synthetic polymer that is much stronger than nylon. In fact, the same weight of Kevlar is five times stronger than steel. It is used in bicycle tires, body armor, and drumheads because of this amazing property. Figure 1.8 shows a model of a segment of Kevlar. Color the O atoms red, the N atoms blue, and the Cl atoms green.

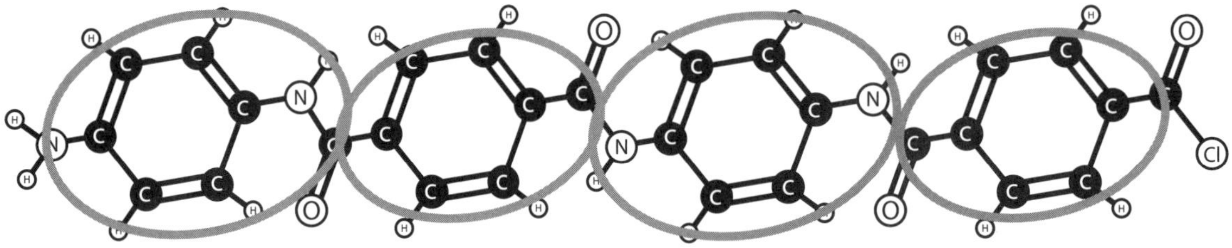

Figure 1.8. Segment of Kevlar Polymer

Figure 1.9 shows the two monomers (para-phenylenediamine and terephthaloyl chloride) that react to form the Kevlar polymer. Color the O atoms red, the N atoms blue, and the Cl atoms green.

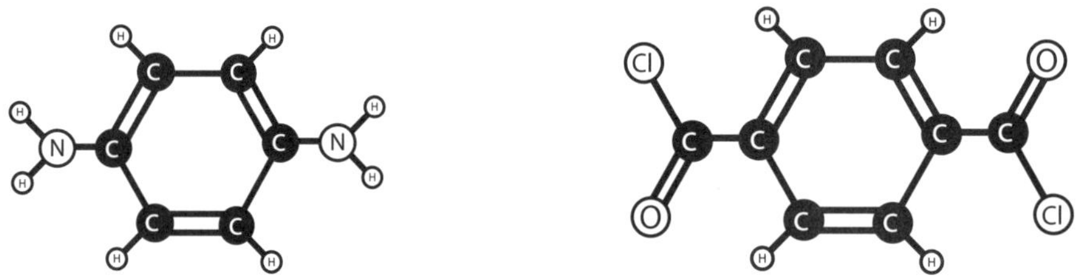

para-phenylenediamine monomer terephthaloyl chloride monomer

Figure 1.9. Monomers Used to Make Kevlar

a. Circle the atoms of the monomers that rearrange to form the Kevlar polymer.

b. Draw models to represent any other products that you predict will form.

Teacher Talk and Actions

Pulling It Together (cont)

Whole class discussion of Pulling It Together Question 3: To see if students are relating phenomena to models.

• Have students look back at the nylon formation reaction and ask whether or not it would be safe for factories to remove the nylon thread and dispose of the other product in a river. (Students should say it's probably OK because the other product is H_2O.)

• Have students look at the Kevlar reaction and ask whether or not it would be safe for factories to remove the Kevlar and dispose of the other product in a river. (Some students may know that HCl is a strong acid, so it would NOT be OK to dispose of it in a river.)

Closure and Link (2 min)

Whole class:

Summarize the lesson:

We used models to help us visualize what happens to atoms and molecules during chemical reactions.

We learned that very large molecules (polymers) can be made by reacting monomers and that water molecules are typically also produced.

We learned that a chemical reaction occurs even when only a few atoms rearrange. You will observe other examples of this in Chapters 3 and 4.

Toward High School Biology

Chapter 2

Teacher Edition

Chapter 2 Overview

To the unaided eye, many chemical reactions that occur in nature give the illusion that matter is either created or destroyed, or just appears or disappears. Why is that? For one thing, such reactions often involve colorless gases whose presence isn't obvious. If chemical reactions occur in open containers the production or consumption of colorless gases may not be noticed. Unless bubbles are produced, gaseous products will not be detected as they leave the reaction container and spread out into the environment. And without measurements, the decrease in the amount of a gaseous reactant will not be detected. In Chapter 1, Lesson 1.2, students used a splint test to "see" that oxygen was consumed when iron rusts and limewater test to detect the production of carbon dioxide gas when baking soda reacted with vinegar. Chapter 2 develops the concept of matter conservation in terms of both atom conservation and mass conservation: **When atoms that make up the molecules of reactants rearrange to form molecules of products during chemical reactions, the number and mass of each type of atom does not change (atoms are conserved) and, therefore, the total mass of substances involved in the reaction does not change (mass is conserved).** Chapter 2 also helps students reconcile how matter can be conserved even if the mass of a system changes: If atoms can enter the system and react to form products or atoms can leave the system as products then the system will weigh more or less than it did at the start. The chapter introduces the term *measured mass* to help students appreciate that a change in measured mass does not violate conservation principles.

Lesson 2.1 provides evidence that when all reactants and products are trapped in a container, measured mass remains constant; however, when the container is opened, the measured mass can change. Students begin to make sense of their observations in terms of all substances being trapped in a sealed container or, in the case of gases, substances being able to enter or leave the opened container. The lesson provides clear evidence that gases have mass and that measured mass does not always reflect the total mass of the substances involved in a chemical reaction.

Lesson 2.2 provides a structured modeling activity to help students visualize that neither the number of atoms nor the mass of atoms changes when atoms rearrange during chemical reactions. By weighing the models of reactants and products, students can see that the total mass does not change when all of the atoms and molecules of reactants and products are accounted for.

Lesson 2.3 uses models to demonstrate why, then, measured mass changes when the containers are opened. By weighing the models of reactants and products, students can see that if a gaseous reactant is limiting, then molecules of gas reactants can enter an opened container and react with remaining molecules of other reactants to form more products, which increases the amount of atoms/matter (hence, measured mass) in the container. And gas products can leave the opened container, which decreases the amount of atoms/matter (hence, measured mass) in the container.

Chapter 2 Background Knowledge for Teachers

The "Three Changes" Revisited

Chapter 2 uses the same three reactions used in Chapter 1 to help students understand that neither the number of atoms nor the mass of those atoms changes when they are rearranged during chemical reactions. Students will be asked to differentiate between total mass (the mass of all substances/atoms involved in the chemical reaction) and the measured mass (the mass of the substances/atoms that are being measured by the balance). When chemical reactions are carried out in sealed containers (closed systems), the total mass is equal to the measured mass, and both remain the same as the reaction proceeds. When containers are opened (open systems), the measured mass may not always reflect the total mass if substances (usually gases) involved in the chemical reaction enter or leave the container (system).

If gaseous reactants enter the system and react to form more products,

- the measured mass of the reactants originally inside the system does not reflect total mass of all reactants (reactants originally inside system plus gas reactants once outside of the system). Thus, the total mass of the reactants is greater than the measured mass of the reactants inside the system.

And when gaseous products leave the system,

- the measured mass of the products left in the system does not reflect total mass of all products (products still inside the system plus gas products now outside of the system). Thus, the total mass of the products is greater than the measured mass of the products inside the system.

Steel Wool (Iron) and Air (Oxygen)

Recall that the reaction of iron in steel wool and oxygen in the air is represented with the chemical equation:

$$\mathbf{4\ Fe\ (s) + 3\ O_2\ (g) \rightarrow 2\ Fe_2O_3\ (s)}$$

Given sufficient time or a catalyst to speed up the reaction, all the iron will react with oxygen to form rust. As this happens, an increase in measured mass occurs from chemically combining iron (present in the container) and oxygen (mostly from outside the container).

Since oxygen atoms have mass, rusted steel wool weighs more than the original steel wool. If the entire sample of steel wool rusted, it would weigh 30 percent more than its original mass (amu stands for atomic mass units):

$4\ Fe\ (s)$ +	$3\ O_2\ (g)$	→	$2\ Fe_2O_3\ (s)$
4 x 56 amu	3 x 32 amu		2 x 160 amu
224 amu	96 amu		320 amu

The video students watch in Lesson 2.1 comes close to this approximation.

Intended observations: Students should observe that when the reaction is carried out in a sealed container, a gas is used (as evidenced by the inverted balloon) and a new solid forms but the measured mass does not change. When the container is opened, the measured mass will increase over time until all of the iron has reacted to form rust.

Rationale for this reaction: Rusting provides an example of a gaseous reactant becoming part of a solid product. In this reaction, the increase in measured mass that occurs when the container is opened is due to the additional oxygen that becomes chemically combined with iron. Because iron is in excess at the beginning of the reaction (i.e., oxygen is limiting), the excess iron is available to react with additional oxygen when the container is opened. Careful observations challenge common misconceptions that matter decreases or is destroyed during rusting (probably based on students having observed the apparent decrease in mass when a car rusts because the rusted product fell off of the car). The rusting reaction is intended to show students that a gaseous reactant that enters an open system can lead to an increase in mass, which will serve as an analogy to the overall reaction for photosynthesis, where carbon dioxide gas combines with water to form solid glucose monomers, which can react to form polymers that make up plant body structures.

Baking Soda and Vinegar

Recall that the reaction of vinegar (acetic acid) and baking soda (sodium bicarbonate) is represented with the following equation:

$$HC_2H_3O_2\ (aq) + NaHCO_3\ (s) \rightarrow NaC_2H_3O_2\ (aq) + H_2O\ (l) + CO_2\ (g)$$

Intended observations: Students should observe that when the reaction is carried out in a sealed container, a gas forms (as evidenced by bubbles) but the measured mass does not change. When the container is opened, the measured mass decreases.

Rationale for this reaction: This reaction produces a gaseous product that is readily detected by the appearance of bubbles and should highlight the need to consider the involvement of gases as either products or reactants. Mass measurements of both the iron rusting reaction and the baking soda and vinegar reaction challenge common misconceptions that gases are not made of matter or do not have mass.

Hexamethylenediamine and Adipic Acid

Recall that the reaction of hexamethylenediamine and adipic acid to produce nylon is represented with the following equation, even though it shows only the formation of a dimer from two monomers:

$$C_6H_{16}N_2 + C_6H_{10}O_4 \rightarrow C_{12}H_{24}O_3N_2 + H_2O$$

Intended observations: Students have already observed that the amount of nylon increases and the amounts of the reactants decreases, but they cannot measure the mass of either reactants or products as the reaction proceeds. Rather, they will apply their developing ideas about conservation and measured mass to predict what they would expect to happen to the mass as nylon forms. They are expected to predict that

- the mass of the liquids in the beaker decreases as more nylon thread is pulled from the beaker (implying that the nylon thread increases in mass), and
- if all of the liquid reactants were converted to nylon thread (and water), the mass of the nylon thread (and water) produced would be the same as the initial mass of the two liquids.

Rationale for this reaction: This type of chemical reaction is key to understanding the building of protein- and cellulose-based body structures through polymerization of monomers. As more monomers are linked together to "grow" the polymer molecule, the polymer molecule increases in mass. The monomers that add matter (and, hence, mass) to a polymer as it forms come from somewhere—from the two liquids in the container for nylon, from amino acid monomers obtained from food for protein synthesis in animals, or from glucose monomers formed during photosynthesis reactions in plants.

Chapter 2 Prerequisite Knowledge for Students

The chapter assumes that students have developed a basic understanding that all matter, including invisible matter such as gases or vapor, is particulate. The chapter assumes that students have developed or are progressing in their development of understanding that atoms of molecules of reactants rearrange to form molecules of products during chemical reactions.

Chapter 2 Common Student Ideas and Misconceptions

Researchers have documented many incorrect ideas that students commonly have about conservation of mass. This document provides a brief summary of most of the more common incorrect ideas, but it is not exhaustive. Listen for these ideas in your class, but also for other variants that your particular students may have. Actual student responses will usually be shorter and less precise than those listed here. The descriptions here are meant to fully explain the nature of the incorrect idea and to highlight what is unsaid as much as what is said.

If it looks like the amount of matter in a system changes, some students may think that the mass of that system has changed even if it is a closed system. Students may say:

- *When a solid substance forms during a chemical reaction, the mass will increase because the solid is new matter that wasn't there before.*
- *When a solid dissolves in a liquid, the mass decreases because the solid disappears.*

Many students have difficulty conserving mass during processes in which gases take place. Students may say:

- *A gas does not have mass so the mass will decrease if a gas is formed during a chemical reaction.*
- *A gas is lighter than a liquid or a solid so the mass decreases when a gas is formed during a chemical reaction.*

Some students may think atoms can be created or destroyed during chemical reactions, which can lead to an increase or decrease in mass. Students may say:

- *The formation of a gas means that new atoms are being produced so the mass will increase.*
- *If a solid substance disappears during a chemical reaction, the mass will decrease because the atoms that make up the solid were destroyed.*
- *The mass of a container stays the same because some atoms were destroyed but an equal number of new atoms were created.*

Students may also think that the mass of an atom can change leading to a change in the mass of the system. Students may say:

- *The measured mass increased because the atoms that make up the substances got heavier.*
- *The measured mass decreased because the mass of the atoms in the container decreased.*

Other students may think that atoms are changed into energy or energy is changed into atoms during a chemical reaction. Students may say:

- *If a chemical reaction gives off heat or light, the mass will decrease because matter was changed into energy.*
- *If energy is added to a system, the mass will increase because the energy will be turned into new atoms.*

Students may be confused about what is conserved during a chemical reaction. Students may say:

- *The number of molecules is always conserved.*
- *The total number of atoms (not the number of each type of atom) is always conserved.*

Lesson Guide
Chapter 2, Lesson 2.1
Chemical Reactions and Mass

Focus: In this lesson, students explore what happens to the amount of matter (mass) during chemical reactions. Students observe that when all reactants and products are trapped in a container, the mass remains constant; however, when the container is opened, the mass can change. Students make sense of these observations in terms of substances being trapped in a sealed container (total mass) and, in the case of gases, entering or leaving the opened container (total mass unequal to measured mass). This lesson challenges the common idea that mass is not conserved during chemical reactions, in particular, during those involving gases.

Key Question: What happens to the amount of matter (mass) when new substances form during chemical reactions?

Target Idea(s) Addressed

- When substances undergo chemical reactions, the amount of matter at the end (the mass of the ending substances and any leftover starting substances) will always be the same as the amount of matter at the beginning (the mass of the starting substances), even though new substances, such as gases, form.
- If the measured mass of a system changes, it is because one or more substances have entered or left the system. If the measured mass increases, one or more substances must have entered the system. If the measured mass decreases, one or more substances must have left the system.

Science Practice(s) Addressed

- Analyze and interpret data to provide evidence for phenomena.
- Apply scientific ideas, principles, and/or evidence to construct an explanation for real-world phenomena.

Materials

Activity 1: Video–*Baking Soda and Vinegar With Mass Measurement*; Video–*Steel Wool (Iron) and Air With Mass Measurement*

Advance Preparation

Decide whether you will demonstrate or show the video for Activity 1, Step 1. Watch the video in preparation if you are conducting the demonstration.

Organize materials in advance so that each set of partners or table teams has materials when class begins.

Phenomena, Data, or Models	Intended Observations	Purpose	Rationale or Notes
Activity 1 Video of two chemical reactions: • Steel wool (iron) is exposed to air (oxygen). • Baking soda is mixed with vinegar. Mass measurements taken for each chemical reaction at three points: • Initial mass of sealed container • Final mass after reaction occurs in sealed container • Final mass after opening the container	When the two reactions occur in sealed containers, the measured mass does not change. When the container is opened, the measured mass may change: • Increase in mass when iron rusts • Decrease in mass when baking soda reacts with vinegar	Students explain decreases/increases in the measured mass when a chemical reaction occurs in an open system by using substance-level ideas. An ideal explanation includes ideas that (a) total mass is conserved during chemical reactions and (b) if the measured mass of a system changes, it is because one or more substances has left or entered the system.	The activity challenges a common idea that mass is not conserved during chemical reactions that involve gases (e.g., if a gas is produced during a chemical reaction that takes place in a closed system, mass will decrease because gases have no/less mass). The rusting video challenges a common idea that rusted iron weighs less than un-rusted iron. Rusting provides an analogy for photosynthesis in plants, helping students understand that gas reactants can yield solid products, thereby leading to an increase in the mass of an open system.
Pulling It Together The formation of nylon, specifically the pulling of nylon thread from the interface of the two liquids in the beaker	Predictions to be made are as follows: • Mass of the substances in the beaker decreases as more nylon thread is pulled out. • Mass of the nylon thread increases as more nylon thread is pulled out. •Mass of the nylon thread and the water formed is the same as the original mass of the two liquids in the beaker.	Students predict the change in measured mass in an open system after the chemical reaction occurs and justify their prediction using substance-level science ideas.	The questions focus students' thinking about what is inside and outside the system.

Lesson 2.1—Chemical Reactions and Mass

What do we know and what are we trying to find out?

When living things grow, they gain weight. For example, a baby weighs about 8 pounds when born. After 6 months, she weighs about twice that—16 pounds. By age 3, she is about 30 pounds. Where does the extra matter come from that makes her weigh more?

In Chapter 1, Lesson 1.5, we saw that new substances are made when atoms that make up the molecules of the starting substances (or reactants) rearrange to form the molecules of the ending substances (or products). Understanding how new substances are formed in simple systems will help us build an explanation of how living things grow and why they gain weight when they grow.

First, let's see if the amount of matter (we call this mass) changes during the chemical reactions we have investigated. To figure this out, we have to keep track of all the matter involved in each chemical reaction. One way to do this is to carry out the reaction in a sealed container so that nothing can enter or leave the container. Then we can open the container and observe if the mass has changed. Later, we will consider what this has to do with the growth of living organisms.

With a partner, think about what happens to the amount of matter during each chemical reaction you have studied so far. Answer the Key Question for one of the chemical reactions listed.

- Steel Wool (Iron) and Air (Oxygen)
- Hexamethylenediamine and Adipic Acid
- Baking Soda and Vinegar

Key Question: What happens to the amount of matter (mass) when new substances form during chemical reactions?

In Chapter 1 we learned that atoms are just rearranged, not created or destroyed, during chemical reactions, so this means that the number of atoms stays the same. I guess if the number of atoms stays the same then the amount of matter (mass) should stay the same. For example, when we used LEGO bricks to model the reaction between iron and oxygen, we saw that the number of iron bricks and oxygen bricks was unchanged. We also observed that the number of each type of LEGO brick didn't change when we modeled the baking soda and vinegar reaction. So the mass of all the bricks shouldn't change.

Teacher Talk and Actions
What do we know and what are we trying to find out? (3 min)
Whole class: Read text under "What do we know and what are we trying to find out?" aloud to the class.
Key Question (5 min)
Individuals: Students respond to the Key Question: **What happens to the amount of matter (mass) when new substances form during chemical reactions?** Whole class: Lead discussion. Invite students to share ideas about the Key Question. The goal is primarily eliciting and probing student ideas here. When students share their answers, do not indicate if their answers are right or wrong. However, do not let students assume by your silence that they were correct. Make it clear that students are expressing ideas, but those ideas may change based on what they will see in the lessons. (Remind students of this regularly.) Students may have ideas that are inconsistent with the atom rearrangement model developed in Chapter 1. Do not correct them. Try to identify if students hold common ideas that matter is created or destroyed during chemical reactions or that gases do not have mass.

Activity 1: Measuring Mass

Your teacher will demonstrate or show a video of the chemical reaction Baking Soda and Vinegar. Then you will watch a video of the Steel Wool (Iron) and Air (Oxygen) reaction. Each reaction will be carried out in a sealed container. A balance will be used to measure the mass before the chemical reaction. Then the container will be opened, and mass will be measured again.

Procedures and Questions

1. Watch the demonstration or video for the Baking Soda and Vinegar chemical reaction, and record your observations in Table 2.1.

Table 2.1. Mass of the Baking Soda and Vinegar Reaction

Initial mass (g) of the reactants in the sealed container	48.67/8	g
Final mass (g) after the reaction occurs in the sealed container	48.66/7	g
Final mass (g) after opening the container	47.32/3	g
What else do you observe (if anything) when the container is opened? A whoosh sound when the bottle is opened.		

2. Answer the questions below based on your observations.

   a. Was the mass that we measured on the scale (measured mass) after the chemical reaction occurred in the sealed container different from the initial mass of the reactants? If so, how was it different?

   Measured mass stayed the same.

   b. Did the measured mass change after the container was opened? If so, how did it change?

   Measured mass decreased.

   c. In previous lessons, you learned that carbon dioxide gas was formed in this reaction. Does carbon dioxide gas have mass? What is your evidence?

   Yes. If the gas didn't have mass, the measured mass of the sealed container would not have decreased. The measured mass didn't change until the container was opened, and the gas could leave.

Teacher Talk and Actions	
Activity 1	**(25 min)**

Whole class:

Emphasize that
• mass is a measure of the amount of matter,
• the video will show mass measurements at three time points, and
• small fluctuations in mass readings (0.01–0.02 g) are typical for scales that do not have shielded pans (moving air creates small fluctuations) and are not considered evidence of mass changing.

Baking soda and vinegar:

• Poll students for predictions: Will mass decrease, increase, or stay the same?
• Show Video—*Baking Soda and Vinegar With Mass Measurement* or demonstrate this (Step 1).
• Students complete Table 2.1 and respond to questions in Step 2.

Whole class:

Lead discussion. Invite students to summarize their observations from Table 2.1 and share ideas about the questions in Step 2.

d. What do you think happened to cause the measured mass on the scale to change after the container was opened?

After the container was opened the gas could leave so there was less matter being measured on the scale.

e. Draw and label a diagram that shows what you think happened to the amount of matter when the container was opened.

An ideal response will show carbon dioxide leaving the opened container. This could be shown as words (carbon dioxide gas), pictures, substance names, or chemical formulas of molecules (CO_2) and arrows showing their movement out of the flask.

3. Watch the demonstration or video for the Steel Wool (Iron) and Air (Oxygen) chemical reaction, and record your observations in Table 2.2.

Table 2.2. Mass of the Steel Wool (Iron) and Air (Oxygen) Reaction

Initial mass (g) of reactants in the sealed container	*139.03/4* g
Final mass (g) after the reaction occurs in the sealed container	*139.03/4* g
Final mass (g) after opening the container	*141.78/9* g
What else do you observe (if anything) when the container is opened? *Only a little rust forms when the flask is sealed with the balloon. More rust forms when the balloon is removed. The rusted steel wool pad has more mass (~2.5 more grams) than the un-rusted steel wool pad.*	

4. Answer the questions below based on your observations.

a. Was the measured mass after the chemical reaction occurred in the sealed container different from the initial mass?

Measured mass stayed the same.

Teacher Talk and Actions

Activity 1 (cont)

Question 2d: What do you think happened to cause the measured mass on the scale to change after the container was opened?

Students should reach consensus that

- carbon dioxide gas has mass,
- carbon dioxide gas left the container when it was opened, and
- the mass decreased because the carbon dioxide gas left the opened container.

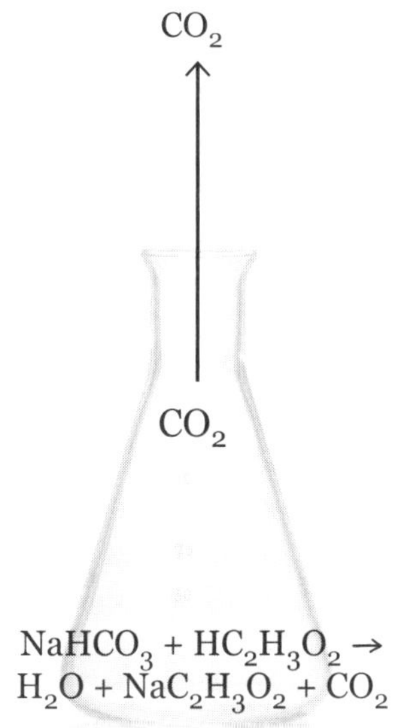

CO_2 gas produced leaves the open flask.

Baking Soda + Vinegar → Water + Sodium Acetate + Carbon Dioxide

Steel wool (iron) and air:

- Poll students for predictions: Will mass decrease, increase, or stay the same?
- Show Video–*Steel Wool (Iron) and Air With Mass and Measurement* (Step 3).
- Students complete Table 2.2 and respond to questions in Step 4.

Whole class:
Lead discussion. Invite students to summarize their observations from Table 2.2 and share ideas about Step 4.

b. Did the measured mass change after the container was opened? If so, how did it change?

Measured mass increased.

c. In previous lessons, you learned that oxygen gas from the air was used in this reaction. Does oxygen gas have mass? What is your evidence?

Yes. If oxygen gas didn't have mass, the measured mass would not have increased when the container was opened and gas could enter and react.

d. What do you think happened to cause the measured mass on the scale to change after the container was opened?

More oxygen came into the flask and reacted with iron to form more rust that we could measure. Once the oxygen atoms were part of Fe_2O_3, which is a solid, they were "trapped" in the container.

e. Draw and label a diagram that shows what you think happened to the amount of matter when the container was opened.

An ideal response will show both oxygen gas entering the opened container and more rust forming. This could be shown as words (oxygen gas), pictures, or chemical formulas of molecules (O_2) with arrows showing movement into the flask. More rust forming could be shown with words or with pictures/formulas of molecules (Fe_2O_3).

Teacher Talk and Actions

Activity 1 (cont)

Question 4d: What do you think happened to cause the measured mass on the scale to change after the container was opened?

Students may reach consensus that

- gases (oxygen in this case) have mass,
- oxygen gas entered the container when it was opened, and
- the mass increased because oxygen gas entered the opened container.

Students may not come to understand that the mass increased because the gas entered the opened container AND THEN reacted with un-rusted steel wool to form more products. Do not correct them at this point, but be sure to challenge this line of thinking with the follow-up questions provided during this lesson as well as during Lessons 2.2 and 2.3.

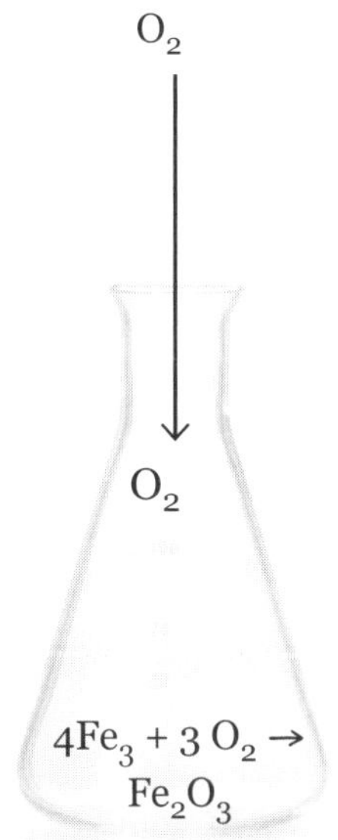

O_2 gas enters the open flask and reacts to make more rust.

Steel Wool + Air → Rust

(Iron) (Oxygen) (Iron oxide)

5. Think about your observations of both chemical reactions. Discuss these questions with your team and record your ideas.

   a. After a gas leaves an opened container, is it still being measured by the scale?

   No

   b. When a gas enters an opened container and reacts with the substances in the container, is it part of the initial measurement of the reactants on the scale?

   No

   c. If the measured mass changes during a chemical reaction, what could be the cause? (Remember, the measured mass is the mass of the amount of matter that is measured by the scale.)

   Whether or not the container is open and whether or not a gas is a reactant or product (a gaseous product could escape and a gaseous reactant could enter)

   d. Does the *total mass* change during a chemical reaction? (The total mass is the mass of *all* matter involved in the chemical reaction—what is measured and any that *is not* measured by the balance.) Why or why not? Explain your answer.

   No, the total mass stays the same. The measured mass did not change after either chemical reaction when the container was sealed. But we are not measuring the mass of all of the reactants or all of the products if some can enter or leave an opened container.

Teacher Talk and Actions

Activity 1 (cont)

Partners:

Students complete Step 5.

Whole class:

- Review the terms in Step 5c and Step 5d:
- The measured mass is the mass of matter involved in the chemical reaction that is measured by the balance.
- The total mass is the mass of all matter involved in the chemical reaction—what is measured and any that is not measured by the balance.
- Poll students for a quick check of students' thinking about Step 5.
- Ask for volunteers of each "side" to explain their thinking.

If the class period is almost over, consider stopping here. If you stop here:

Have students summarize their learning with Pulling It Together Question 1.

Assign Pulling It Together Questions 2 and 3 for homework, with discussion to follow in class tomorrow.

Pulling It Together

Work on your own to answer these questions. Be prepared for a class discussion.

1. Given your observations, how would you answer the Key Question, **What happens to the amount of matter (mass) when new substances form during chemical reactions?**

The amount of matter (mass) doesn't change in a closed system. However, the measured mass can change if atoms/matter (mass) are allowed to enter or leave the system. This can happen if a gas is either a reactant or product of a chemical reaction. If a gas produced during the reaction leaves, the mass decreases and if a gas enters and reacts, the mass increases.

2. Think back to the video that shows how nylon forms. Nylon is made from two liquids—hexamethylenediamine and adipic acid. Nylon and water form where the two liquids touch each other.

   a. What do you think will happen to the measured mass of the nylon as more and more of it forms?

   The measured mass of the substances in the beaker would decrease because you remove a substance (the nylon product) from the container. Water is also a product of the reaction and some of it might be removed with the nylon. (The nylon looked moist, so maybe water was stuck to it.)

   b. What will happen to the measured mass of the liquids in the beaker as more and more nylon thread is pulled out? Why do you think so?

   The measured mass of the liquids in the beaker would decrease as more nylon thread is removed from the beaker. Atoms making up the molecules of the monomers (hexamethylenediamine and adipic acid) are rearranged to form nylon polymer and water molecules. As nylon (and water) is removed there is less matter in the beaker, which means there is less mass.

   c. Do you think the measured mass of all the nylon that could be formed would be less than, equal to, or more than the original measured mass of the two liquids? Why do you think so?

   Water is also formed during the chemical reaction that makes nylon. The total mass of all the nylon and water produced would be equal to the original mass of the two liquids (hexamethylenediamine and adipic acid). Just the mass of the nylon would be less than the mass of the two liquids.

Teacher Talk and Actions	
Pulling It Together	**(15 min)**

Individuals:

Students answer Pulling It Together questions on their own, in detail.

Whole class:

Lead discussion. Invite students to share ideas about Pulling It Together Question 1 (the Key Question) and Question 2c.

Direct students to look back at their initial ideas about the Key Question and consider how their ideas have changed.

3. Given what you learned in previous lessons about chemical reactions, how do you think atoms and molecules could be used to explain the following observations about mass?

   a. Measured mass stays the same when a chemical reaction takes place in a container that is sealed.

   If the container is sealed atoms/molecules can't enter or leave, so the number of atoms in the container stays the same.

   b. Measured mass sometimes changes when a chemical reaction takes place in a container that is open.

   Atoms/molecules of reactants or products (especially gases) can enter or leave if the container is opened. Because atoms have mass, the measured mass changes.

SE p. 76

Teacher Talk and Actions

Pulling It Together (cont)

Pulling It Together Question 3. In Lesson 2.2, students will be introduced to the idea that the mass of an atom doesn't change in a chemical reaction (Science Idea #8). You could foreshadow the idea for your more able students by asking what would happen if the mass of an atom could change during a chemical reaction. Students could reason that the mass of individual LEGO bricks didn't change when they modeled chemical reactions, so perhaps the mass of real atoms don't change.

Closure and Link (2 min)

Whole class:

Summarize the lesson:

The total amount of matter (total mass) stays constant during chemical reactions. We observed that mass didn't change when we trapped all starting and ending substances in the container.

The measured mass can change when starting or ending substances (usually a gas) can enter/ leave an opened container.

Link to the next lesson:

In the next two lessons,we will use models to help us think about what is happening in both cases to ensure we all understand the difference between total mass and measured mass and sealed containers and opened containers. You can test the ideas you wrote down in Pulling It Together Question 3.

Lesson Guide
Chapter 2, Lesson 2.2
Sealed Containers and Total Mass

Focus: In this lesson, students use LEGO models to visualize that neither the number of atoms nor the mass of atoms changes when they are rearranged during chemical reactions. The models demonstrate that mass does not change when all of the atoms and molecules of reactants and products are accounted for (referred to as total mass). This lesson challenges the common idea that mass is not conserved during chemical reactions, in particular, during those involving gases.

Key Question: Why does rearranging atoms keep the *total* mass constant during chemical reactions?

Target Idea(s) Addressed

- Atoms are not created or destroyed during chemical reactions, so the total number of each type of atom remains the same.
- The mass of a particular atom does not change during chemical reactions, so a given number of that type of atom will always have the same total mass.
- Because the mass of a particular atom does not change and because the number of each type of atom stays the same, the total mass of the matter stays the same, even though the atoms are rearranged during a chemical reaction.

Science Practice(s) Addressed

- Use a model to describe phenomena.
- Apply scientific ideas, principles, or evidence to construct, revise, or use an explanation for real-world phenomena, examples, or events.

Materials

Activity 1: Copymaster: *Model Key,* LEGO model kit

Activity 2: Copymasters: *Model Key, Chemical Reaction Mats:* Steel Wool (Iron) and Air (Oxygen), and Baking Soda and Vinegar; LEGO model kit, Electronic scales, Plastic zipper bags

Advance Preparation

Organize materials in advance so that each set of partners or table teams has materials when class begins.

Phenomena, Data, or Models	Intended Observations	Purpose	Rationale or Notes
Activity 1 LEGO bricks and atomic mass data for H, C, N, and O atoms	LEGO bricks have mass that is dependent on brick size. The mass of a LEGO brick does not change regardless of how it is arranged with other bricks or what other brick it is linked to.	Students consider how the mass of LEGO bricks can be used to represent relative atom masses.	The activity challenges common ideas that atoms do not have mass. The analogy between LEGO brick size (mass) and atom mass is not a perfect one. The claim that atoms have discrete masses that LEGO size (mass) can approximate works primarily for H, C, N, and O. The ratio of LEGO brick size to atom mass is not accurate.
Activity 2 LEGO models representing two chemical reactions: • Steel wool (iron) and air (oxygen) react to form rust. • Baking soda and vinegar react to form carbon dioxide, sodium acetate, and water.	When modeling the chemical reactions, the number of each type of LEGO brick is the same in the reactants and in the products. When modeling the chemical reactions in sealed containers, the mass of LEGO reactant models is the same as the mass of LEGO product models.	From two examples, students generalize the principle/idea that mass remains constant during a chemical reaction in which all of the reactants and products are trapped because (a) atoms aren't created or destroyed; (b) since atoms didn't enter or leave, the system has the same number of each type of atom at the end of the reaction as at the start; and (c) the mass of a particular atom does not change, so the same number of that type of atom will always have the same mass.	The activity provides an opportunity for students to see an example of the crosscutting concept of matter conservation, which will be extended in Lesson 2.3 and connected to animal and plant growth in Lessons 3.4, 4.3, and 4.5.
Pulling It Together LEGO models are used to show what leaves the open container when baking soda and vinegar react.	If LEGO models of CO_2 are removed from the balance, the measured mass will decrease. If we use the mass of LEGO bricks, we can estimate the change in measured mass.	Students use LEGO bricks to reason about changes in measured mass during chemical reactions in open systems.	The activity provides an opportunity for students still struggling to "think with" a particle model to visualize the link between mass conservation and atom conservation.

Lesson 2.2—Sealed Containers and Total Mass

What do we know and what are we trying to find out?

We saw in Lesson 2.1 that the measured mass does not change when we carry out a chemical reaction in a sealed container even though the products (ending substances) are different from the reactants (starting substances). Because nothing can enter or leave the sealed container, the total amount of the matter of all the substances does not change during chemical reactions that occur in sealed containers.

Remember that all matter is made up of atoms, and we learned in the earlier lessons that atoms are rearranged—not created or destroyed—during chemical reactions. This means that the number of each type of atom stays the same during chemical reactions. In this lesson, you will use models to help you think about what atoms have to do with our observations of mass during chemical reactions that take place in sealed containers.

Think about your experiences representing chemical reactions with models. Then respond to the Key Question to the best of your knowledge.

Key Question: Why does rearranging atoms keep the *total* mass constant during chemical reactions?

In Chapter 1 we learned that atoms are just rearranged, not created or destroyed, during chemical reactions. We saw that the number of LEGO bricks stayed the same, so this suggests that the number of atoms stays the same. Because atoms are what gives matter its mass, if the number of atoms stays the same then the total mass would stay the same.

Teacher Talk and Actions

What do we know and what are we trying to find out? (3 min)

Whole class:

Read text under "What do we know and what are we trying to find out?" aloud to the class.

Key Question (5 min)

Individuals:

Students respond to the Key Question: **Why does rearranging atoms keep the *total* mass constant during chemical reactions?**

Whole class:

Lead discussion. Invite students to share ideas about the Key Question.

The goal is primarily eliciting and probing student ideas here.

When students share their answers, do not indicate if their answers are right or wrong. However, do not let students assume by your silence that they were correct. Make it clear that students are expressing ideas, but those ideas may change based on what they will see in the lessons. (Remind students of this regularly.)

Activity 1: Comparing the Mass of LEGO Bricks and Atoms

Materials
For each team of students
LEGO model kit
Model Key

In this activity, you will use the mass of LEGO bricks to help you think about the mass of the atoms they represent.

Procedures and Questions

1. Every type of atom has a unique mass. Scientists have determined the mass of each type of atom. Examine Table 2.3 for the masses of hydrogen, carbon, nitrogen, and oxygen, given in atomic mass units (amu).

Table 2.3. Atom Masses

Atom	Approximate Mass (amu)
Hydrogen	1
Carbon	12
Nitrogen	14
Oxygen	16

2. Discuss the following questions with your team and record your answers.

   a. What do you notice about the mass of a hydrogen atom in comparison with that of a carbon atom?

   Hydrogen atoms have less than 1/10 the mass of carbon atoms.

   b. What do you notice about the mass of a carbon atom in comparison with that of a nitrogen atom or an oxygen atom?

   The masses of carbon, nitrogen, and oxygen atoms are similar.

Teacher Talk and Actions	
Activity 1	**(10 min)**
Small group/partners: Students complete Steps 1–3. Alternatively, lead a whole class discussion of Steps 1–3.	

3. With your team, contrast the masses of these LEGO bricks representing different atoms:

   a. A white brick (hydrogen atom) and a black brick (carbon atom)

   White LEGO bricks have less mass. They are smaller in size than black bricks. According to the table the hydrogen atoms also have less mass, so maybe they are smaller in size than carbon atoms, too.

   b. A black brick (carbon atom), a blue brick (nitrogen atom), and a red brick (oxygen atom)

   Black, blue, and red LEGO bricks have the same mass. They are the same size. According to the table, the carbon, nitrogen, and oxygen atoms have similar mass and size.

4. Discuss the following questions with your team and record your answers.

   a. Since there are only two different sizes of LEGO bricks in your kit, how could you use LEGO models to represent four different atoms?

   We could use the small brick to represent the smallest type of atom and use the large brick to represent the other 3 types of atoms because they are more similar in mass. We could use different colored bricks to distinguish among the 3 types. That would allow us to use models to show atom rearrangement.

   b. If you were to design a custom LEGO kit, how many different size bricks would you use to represent the four types of atoms in Table 2.3, and what would their relative masses be?

   Because data in Table 2.3 shows that each type of atom has a different mass, we would need 4 different size bricks to represent the mass of the 4 different types of atom. The brick representing a carbon atom would be 12 times as heavy as the brick representing a hydrogen atom, the brick representing a nitrogen atom would be 14 times as heavy as the brick representing a hydrogen atom, and the brick representing an oxygen atom would be 16 times as heavy as the brick representing a hydrogen atom.

5. Discuss the following questions with your team and record your answers.

   a. In Lessons 1.5 and 1.6, when you built LEGO models of reactant molecules and rearranged the bricks to form product molecules, did the mass of any individual brick change?

   No, the mass of each brick stayed the same.

   b. Do you think the mass of an actual atom changes during a chemical reaction? Does the mass of an atom change depending on which other atoms it is connected to? Why or why not?

   Probably not. Our models showed that even when bricks (atoms) were attached to different atoms (like oxygen was first connected to oxygen, then it was connected to iron), the mass of the brick didn't change.

Teacher Talk and Actions
Activity 1 (cont)
Small group/partners: Students complete Steps 3, 4, and 5. Alternatively, lead a whole class discussion. Students should reach consensus that atoms have different masses and that LEGO bricks can represent approximate masses. Step 5. You may be surprised by your students' thinking. You may hear things such as "The mass of a LEGO brick can change. When it's on the top of a pile, it has more force pushing down on it, so it will be heavier." Or " I think the LEGO brick can get lighter when it's on top of a big pile of atoms. The scale can't "feel it" as much." We have heard these ideas before in field-test classrooms. While uncommon, these ideas indicate some interesting ways in which students try to make sense of position, gravity, and mass. Do not correct them at this point. These ideas will be challenged in Activity 2.

Activity 2: Using Models to Think About Chemical Reactions in Sealed Containers

Materials
For each team of students
LEGO model kit
Chemical Reaction Mats: Steel Wool (Iron) and Air (Oxygen), Baking Soda and Vinegar
Model Key
Access to an electronic scale
Plastic zipper bag

In this activity, you will use models to help you think about what atom rearrangement during chemical reactions has to do with the mass of sealed containers staying constant. Each LEGO brick represents one atom.

Remember, the *Chemical Reaction Mats* are one way to represent the starting and ending substances. They show the following:

Reactants (starting substances) ⟶ Products (ending substances)

The reactants are shown on the green side of the mat. The products are shown on the yellow side of the mat.

Procedures and Questions

1. Use the *Chemical Reaction Mats* and LEGO model kit to complete these steps for the Steel Wool (Iron) and Air (Oxygen) reaction. Fill in Table 2.4 as you work.
   a. In the first row of Table 2.4, write a word equation that represents the Steel Wool (Iron) and Air (Oxygen) reaction.
   b. Use LEGO bricks to build the models of reactants as shown on the green side of the mat. Put all other LEGO bricks aside.
   c. Put the models in the plastic bag, seal it, measure the mass of the models of the reactants with a balance, and record the mass in Table 2.4.
   d. Use only bricks inside the bag to build the models of products as shown on the yellow side of the mat.
   e. Seal the bag again, measure the mass of the models of the products with a balance, and record the mass in Table 2.4.

Table 2.4. Steel Wool (Iron) and Air (Oxygen) Reaction

Reactants ⟶ Products	
Equation: iron + oxygen gas → iron oxide (rust) $4\ Fe + 3\ O_2 \rightarrow 2\ Fe_2O_3$	
Mass (g) of the models of the reactants 22 g	Mass (g) of the models of the products 22 g

Teacher Talk and Actions

Activity 2 (15 min)

Emphasize that *Chemical Reaction Mats* represent a chemical reaction occurring in a sealed container. We are accounting for all of the starting and ending substances (and their atoms).

Provide as much or as little up-front "modeling" as you think is necessary to support partners in completing the activity successfully.

Demonstrate proper use of the electronic scales:
- Students should use the same balance when measuring the reactants and the products.
- Tare/zero before each use.

Partners:

Students complete Step 1 for steel wool (iron) in air and Step 2 for baking soda and vinegar.

Monitor partners to ensure that
- all students are engaged in building models;
- models are built accurately;
- once the starting substance models are built, all other LEGO bricks are put away; and
- students take time to think about each chemical reaction after they represent it with models.

2. Repeat Step 1a–e for the Baking Soda and Vinegar reaction. Fill in Table 2.5 as you work.

Table 2.5. Baking Soda and Vinegar Reaction

Reactants ⟶ Products	
Equation: *sodium bicarbonate + acetic acid → sodium acetate + carbon dioxide + water* $NaHCO_3 + HC_2H_3O_2 \rightarrow NaC_2H_3O_2 + CO_2 + H_2O$	
Mass (g) of the models of the reactants	Mass (g) of the models of the products
24 g	*24* g

3. Think about your experience with the models of both chemical reactions. Discuss the following questions with your team and write down your ideas.

   a. For the Steel Wool (Iron) and Air (Oxygen) reaction, what do you notice about the mass of the products in comparison with the mass of the reactants?

   The mass of the models of reactant molecules (4 Fe + 3 O_2) is the same as the mass of the product models (2 Fe_2O_3). The total mass of the models doesn't change.

   b. Look back at Lesson 2.1 and see what you wrote in the first two rows of Table 2.2. Use your answer to Question 3a above and your observations of the models to explain the data you entered in Table 2.2.

   In the sealed container, the initial mass (139.03 g) of reactants is the same as the final mass (139.03 g) because the number of Fe and O atoms is the same. We could show this with models: If we weighed the LEGO models before and after the reaction in a closed container, the total mass would be the same.

   c. For the Baking Soda and Vinegar reaction, what do you notice about the mass of the products in comparison with the mass of the reactants?

   The mass of the models of reactant molecules ($NaHCO_3 + HC_2H_3O_2$) is the same as the mass of the product models ($NaC_2H_3O_2 + CO_2 + H_2O$). The total mass of the models doesn't change.

   d. Look back at Lesson 2.1 and see what you wrote in the first two rows of Table 2.1. Use your answer to question 3c above and your observations of the models to explain the data you entered in Table 2.1.

   In the sealed container, the initial mass (48.67 g) of reactants is the same as the final mass (48.67 g) because the number of Na, H, C and O atoms is the same. We could show this with models: If we weighed the LEGO models before and after the reaction in a closed container, the total mass would be the same.

   e. Write a "rule" or general statement that summarizes your findings recorded in Questions 3a–d. Be sure to use ideas about mass and atoms.

   The total mass doesn't change when a chemical reaction occurs in a closed container, because the total number of each type of atom doesn't change.

Teacher Talk and Actions
Activity 2 (cont)
Partners: Students complete Step 2 using Table 2.5. Whole class: Chart while two teams share their data in Tables 2.4 and 2.5. Other groups should verify findings against their own tables. Notable discrepancies (1 g or more) in mass between tables or between reactants and products are likely to be the result of inaccurately built models. If a table reports an increase or decrease in mass measurements from reactants to products, suggest that students conduct a talk-aloud while you model the chemical reaction again for the class. Students should reach consensus that when modeling the chemical reactions: • the mass of a LEGO brick (atom) does not change, regardless of which atoms it connects to, • the number of each type of LEGO brick (atom) is the same in the reactants and in the products, and • the mass of LEGO models of reactant molecules is the same as the mass of LEGO models of product molecules.

Science Ideas

The activities in this lesson were intended to help you construct important ideas about the relationships between atoms and total mass. Read the ideas below. Remember that because science ideas are consistent with evidence from a wide range of phenomena, you can apply them to observations and data about similar phenomena. Look back through Lesson 2.2. In the space provided after each science idea, give evidence that supports the idea.

Science Idea #7: Atoms are not created or destroyed during chemical reactions, so the total number of each type of atom remains the same. We say that *atoms are conserved.*

Evidence:

When we model chemical reactions according to the Chemical Reaction Mats, the LEGO bricks making up the reactant molecules are just rearranged to form the product molecules. We can see that there is the same number of each type of brick in the product molecules as in the reactant molecules. For example, for the reaction $4\ Fe + 3\ O_2 \rightarrow 2\ Fe_2O_3$ there are 4 bricks representing Fe atoms and 6 bricks representing O atoms in the reactants and 4 bricks of Fe atoms and 6 bricks of O atoms in the products.

Science Idea #8: The mass of a particular atom does not change, so a given number of that type of atom will always have the same *total* mass.

Evidence:

The mass of a red or white or black LEGO brick doesn't change regardless of what other bricks it is connected to.

Science Idea #9: Because the mass of a particular atom does not change and because the number of each type of atom stays the same, the *total mass* of the matter stays the same even though atoms are rearranged during chemical reactions. We say that *mass is conserved.*

Evidence:

We used models to help us think about atoms rearranging during chemical reactions and what that has to do with mass measurements. Since atoms are just rearranged, the number of each kind of atom is the same in the reactants and products. For example, there were 4 Fe atoms and 6 O atoms in the reactants and in the products of the rusting chemical reaction. These atoms have the same total mass, no matter what other atoms they are connected to, so the total mass of 4 Fe bricks and 6 O bricks is conserved.

Teacher Talk and Actions

Science Ideas (5 min)

Whole class:

Tell students: You have used models to help clarify your thinking about the relationship between atoms and mass.

Read the science idea(s) aloud to the class. Add to the list of charted science ideas. Give students time to write down at least one piece of evidence that supports the idea.

When discussing Science Idea #8, ask, "What do you think would happen to the total mass if the mass of an atom could change?" Students may assume that this idea is correct but be unaware of their assumption. Asking this question may help to make the assumption explicit.

Remind students that the science ideas will be used in constructing scientific explanations about similar phenomena.

Pulling It Together

Work on your own to answer these questions. Be prepared for a class discussion.

1. After learning about the relationship between atoms and mass, how would you answer the Key Question, **Why does rearranging atoms keep the *total* mass constant during chemical reactions?** Explain it as you would to a friend who missed class.

   During chemical reactions, the products are made from the same atoms as reactants. The atoms of reactants just rearrange into new combinations in the products. The number of atoms and the mass of atoms do not change when they are rearranged. The models show that total mass does not change when all of the atoms making up the molecules of reactants and products are measured.

2. In Lesson 2.1, you observed that measured mass remained constant after baking soda and vinegar reacted in a sealed container. However, the measured mass decreased when the container was opened.

   When you measured the mass of the LEGO models shown on the *Chemical Reaction Mats*, you measured the mass of all of the models of reactants and all of the models of products. This is like measuring the mass of reactants and products of a chemical reaction in a sealed container. Now, let's use the LEGO models to help us think about what happened to the mass when the container was opened.

Teacher Talk and Actions	
Pulling It Together	**(15 min)**

Individuals:
Students answer Pulling It Together questions on their own, in detail.

Whole class:
Lead a discussion of Pulling It Together Question 1:

- Invite students to share their ideas.

- Direct students to look back at their initial ideas about the Key Question and consider how their ideas have changed.

Note: Pulling It Together Question 2 provides an opportunity for you to see how well students are able to go back and forth between an observable phenomenon (decrease in measured mass) and a model of what happens to molecules (LEGO models of CO_2 molecules leave the container).

As you circulate, check to see if students are able to relate what they actually observed to what they could not observe (and hence had to get ideas from what happens to the models).

a. Look at the *Chemical Reaction Mat* for Baking Soda and Vinegar (also shown in Figure 2.1). Circle the model in Figure 2.1 that represents a gas molecule that could leave the opened container.

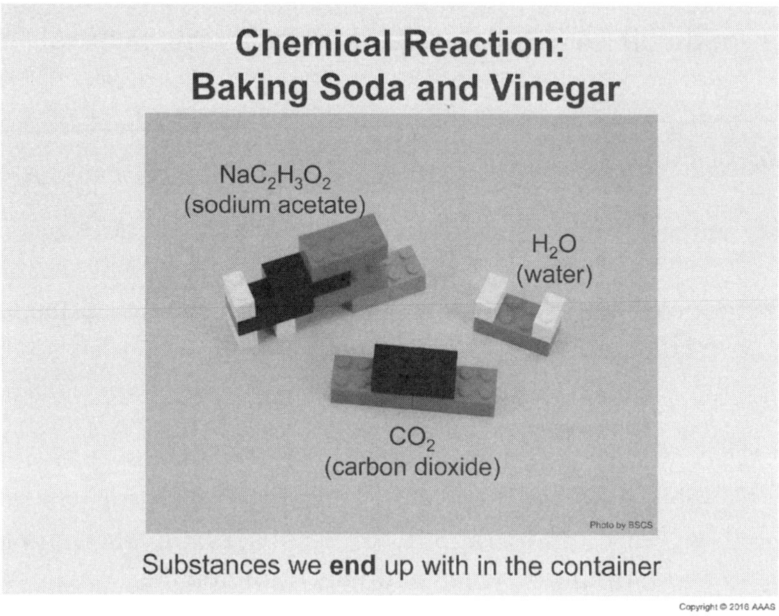

Figure 2.1. *Chemical Reaction Mat*

b. Think about Activity 2, Step 1e, where you measured and recorded the mass of the models of the products with a balance. Imagine that models of all three products (sodium acetate, water, and carbon dioxide) are sitting on a balance.

Describe how you could represent a gas molecule leaving the opened container. Be as specific as you can about which atoms and how many of them will "leave."

During chemical reactions, the products are made from the same atoms as reactants but are rearranged in new combinations: The number of atoms and the mass of atoms do not change when they are rearranged. But, if the carbon dioxide molecules could escape, we would take those off the balance. Then the measured mass of the models would not include models of the carbon dioxide molecules. So, the measured mass of the product models (sodium acetate and water molecules) would be less than the measured mass of the reactant models (baking soda and vinegar molecules).

Teacher Talk and Actions
Pulling It Together (cont)
(Intentionally left blank)

c. Table 2.6 shows how many bricks are measured by the balance and the mass of these bricks before the container is opened. Complete the table to show how many bricks are measured by the balance and the mass of these bricks after the gas molecule leaves the opened container.

Table 2.6. Measured Mass of Molecular Models

	Sealed Container		Opened Container, After Gas Leaves	
Bricks, Mass (g)	**# of Bricks on the Balance**	**Mass of Bricks on the Balance (g)**	**# of Bricks on the Balance**	**Mass of Bricks on the Balance (g)**
Hydrogen brick, 0.5 g	5	5 bricks x 0.5 g/brick = 2.5 g	5	5 bricks x 0.5 g/brick = 2.5 g
Carbon brick, 2 g	3	3 bricks x 2 g/brick = 6 g	2	2 bricks x 2 g/brick = 4 g
Oxygen brick, 2 g	5	5 bricks x 2 g/brick = 10 g	3	3 bricks x 2 g/brick = 6 g
Sodium brick, 2 g	1	1 brick x 2 g/brick = 2 g	1	1 bricks x 2 g/brick = 2 g
Total	14	2.5 g + 6 g + 10 g + 2 g = 20.5 g	11	2.5 g + 4 g + 6 g + 2 g = 14.5 g

d. Did you see a change in the mass of the bricks in the opened container? Which bricks are now gone?

Yes, the total mass of the bricks in the opened container decreases. 1 carbon and 2 oxygen bricks are now gone. This is because a CO_2 left the container.

e. If so, does the amount of the change in mass correspond to the mass of the model you circled in the picture in Step 2a? Show your work.

The carbon dioxide molecule is the one I circled on the previous page. It weighs 6 g, which is the difference between the sealed container and the opened container.

Teacher Talk and Actions
Closure and Link **(2 min)**
Whole class: Summarize the lesson: The *Chemical Reaction Mats* represent chemical reactions occurring in sealed containers. What would change if we showed what happens when we open the container? We will explore this more in the next lesson. Link to next lesson: What ideas do you have? Lead discussion. Invite students to share ideas about Pulling It Together Question 2 to foreshadow the next lesson. Pulling It Together Question 2: In Lesson 2.1, you observed that the mass of the container decreased when it was opened. Think about how you could use atoms and molecules to explain the change in mass.

Lesson Guide
Chapter 2, Lesson 2.3
Opened Containers and Measured Mass

Focus: Students use LEGO models to demonstrate that, although the type and number of each type of atom (hence, total mass) is conserved, measured mass can change when the containers are opened. The models demonstrate that gas products (which are made up of atoms) can leave the opened container, which decreases the amount of matter (hence, measured mass) in the container. The models demonstrate that molecules of gas reactants (also made up of atoms) can enter the opened container and react to form more products, which increases the amount of matter (hence, measured mass) in the container.

Key Question: If atoms and total mass are always conserved during chemical reactions, why can measured mass change when the container is opened?

Target Idea(s) Addressed

- Because the mass of a particular atom does not change and because the number of each type of atom stays the same, the total mass of the matter stays the same, even though the atoms are rearranged during a chemical reaction.
- The measured mass of reactants and products is not always the same as the total mass. The measured mass changes if substances (often gases) enter or leave the system. This is because atoms that make up these substances enter or leave the system.

Science Practice(s) Addressed

- Use a model to describe phenomena.
- Apply scientific ideas, principles, or evidence to construct, revise, or use an explanation for real-world phenomena, examples, or events.

Materials

Activity 2: 1-gallon zipper plastic bag, LEGO model kit, Copymasters: *Chemical Reaction Mats:* Steel Wool (Iron) and Air (Oxygen) and Baking Soda and Vinegar; electronic scales, Video—*Representing Baking Soda and Vinegar With LEGO Models in Sealed and Opened Container With Mass* (optional)

Advance Preparation

Decide whether you will demonstrate or show the video for Activity 2, Step 1. Watch the video in preparation if you are conducting the demonstration.

Organize materials in advance so that each set of partners or table teams has materials when class begins.

Phenomena, Data, or Models	Intended Observations	Purpose	Rationale or Notes
Activity 1 Space-filling models representing a system of hydrogen gas and oxygen gas molecules chemically reacting to form water molecules (and leftover reactant molecules)	When representing chemical reactions with models, some of the reactants may be used up, and some may be left over. The number of molecules is not necessarily conserved during chemical reactions. The number of each type of atom is conserved, hence the total mass of those atoms, during chemical reactions.	Students explain why the total mass is conserved during chemical reactions. An ideal explanation should include the ideas that (a) atoms aren't created or destroyed, though the number of molecules may change, and (b) the mass of a particular atom does not change, so the same number of that type of atom will always have the same mass.	The activity challenges the common idea that conservation of matter applies to both conserving atoms and conserving molecules. The activity provides students with the first opportunity to examine a "system" of reactants and products in which some of the reactants may be used up and some may be left over.
Activity 2 LEGO models of chemical reactions in opened containers (systems): • Baking soda and vinegar • Steel wool (iron) and air (oxygen)	In a sealed container, the number of each type of LEGO brick is the same before and after the chemical reaction. In a sealed container, the total mass of the LEGO bricks is the same before and after the chemical reaction. When LEGO models of product molecules leave an opened container, the number of atoms decreases and the measured mass decreases. When LEGO models of reactant molecules enter an opened container and form new products, the number of atoms increases and the measured mass increases.	Students explain decreases/ increases in the measured mass when a chemical reaction occurs in an open system using atomic-level ideas. An ideal explanation includes the ideas that (a) atoms aren't created or destroyed; (b) the mass of a particular atom does not change, so the same number or type of atom will always have the same mass; (c) therefore total mass is conserved; and (d) if the measured mass of a system changes, it is because atoms have entered or left the system.	The activity provides students with another opportunity to examine a "system" of reactants and products in which some of the reactants may be used up and some may be left over. In the case of rusting, it challenges the common idea that rusted iron weighs less than un-rusted iron.
Pulling It Together The formation of patina on the Statue of Liberty, specifically the production of a layer of copper carbonate on the copper exterior of the statue	(Evidence provided to students) A layer of copper carbonate forms on the statue when copper interacts with oxygen, carbon dioxide, and water vapor in the air.	Students predict the change in measured mass in an open system after the chemical reaction occurs and support their prediction using science ideas about conservation.	The context challenges a common idea that oxidation (e.g., rusting, tarnishing, patina) causes decreases in mass. Patina provides an analogy for photosynthesis in plants, helping students understand that invisible reactants can yield solid products, thereby leading to an increase in the mass of an open system.

Lesson 2.3—Opened Containers and Measured Mass

What do we know and what are we trying to find out?

The *Chemical Reaction Mats* and other models you used in previous lessons helped you think about what happens when all the reactant molecules are used to form product molecules without any leftover reactants. In the real world, however, there are often more molecules of one reactant than another. When this happens, the chemical reaction will continue until molecules of one of the reactants are used up. Then the reaction will stop.

The *Chemical Reaction Mats* also represent chemical reactions as though all reactants and products are trapped in sealed containers. But you saw in Lessons 2.1 and 2.2 that the measured mass *can* change if the container is opened so that substances can enter or leave.

In this lesson, we will use models to help us think about why measured mass changed in the iron-rusting and baking-soda-plus-vinegar reactions after the containers were opened.

At the end of Lesson 2.2, you used LEGO models to think about how rearranging atoms during chemical reactions conserves atoms and total mass. Use your observations from the models to respond to the Key Question to the best of your knowledge. Write about what would cause measured mass to increase and what would cause measured mass to decrease.

Key Question: If atoms and total mass are always conserved during chemical reactions, why can measured mass change when the container is opened?

If a gas entered the container and reacted to form more products, that adds matter to the container and therefore mass. If a gas is produced and leaves the container and is no longer measured, then there is less matter in the container and therefore less mass is measured.

Teacher Talk and Actions

What do we know and what are we trying to find out? (3 min)

Whole class:

Read text under "What do we know and what are we trying to find out?" aloud to the class.

Key Question (5 min)

Individuals:

Students respond to the Key Question: **If atoms and total mass are always conserved during chemical reactions, why can measured mass change when the container is opened?**

Whole class:

Lead discussion. Invite students to share ideas about the Key Question.

The goal is primarily eliciting and probing student ideas here.

When students share their answers, do not indicate if their answers are right or wrong. However, do not let students assume by your silence that they were correct. Make it clear that students are expressing ideas, but those ideas may change based on what they will see in the lessons. (Remind students of this regularly.)

Do not correct students if they seem to be struggling with measured mass versus total mass. Activity 2 will help them visualize when they are the same and when (and why) they are different.

Activity 1: Reactants, Products, and Leftovers

Figure 2.2 shows what happens when 10 H_2 molecules and 10 O_2 molecules (left) react to form water molecules (right). You will use this image to help you think about what happens when many reactants are available to undergo a chemical reaction.

Using chemical formulas, the equation for making water from hydrogen gas and oxygen gas is as follows:

$$2\,H_2 + 1\,O_2 \longrightarrow 2\,H_2O$$

In this equation, hydrogen molecules (H_2) and the oxygen molecule (O_2) are shown on the left side of the arrow because they are reactants (starting substances). Water molecules (H_2O) are shown to the right of the arrow because they are the products (ending substance). The equation shows the smallest whole number ratio of reactants and products. That is, when two hydrogen molecules react with one oxygen molecule, two water molecules are made. If four hydrogen molecules react with two oxygen molecules, four water molecules are made, and so forth.

Procedures and Questions

1. Look carefully at Figure 2.2, or at a color version your teacher projects and discuss with your team what you see in the image.

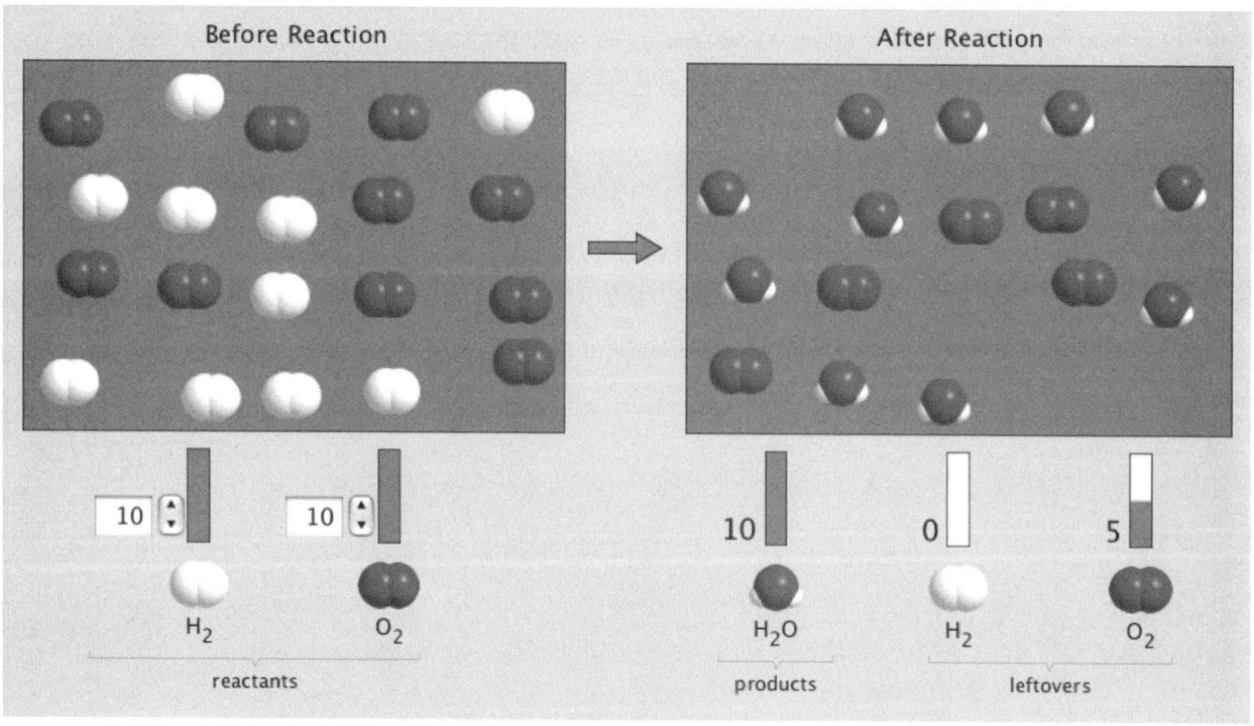

Figure 2.2. Reactants and Products, Before and After Reaction

Teacher Talk and Actions

Activity 1 **(15 min)**

Whole class:

Read introductory text.

The equation for forming water is to help students understand that fixed amounts of reactants interact to form products. Don't digress here to teach about balanced equations.

Whole class:

Lead discussion. Invite students to share ideas about Figure 2.2 (Step 1).

Alternatively, project the PhET Reactants, Products, and Leftovers simulator and run the same example provided in the text. You will need a recent version of Java to run this simulation:

1) Go to the simulator URL:
http://phet.colorado.edu/en/simulation/reactants-products-and-leftovers.
2) Select **Run Now!** and open simulator.
3) Select **Real Reaction** tab.
4) Select **Make Water**.
5) Enter the reactants as shown in the text and run.

Science Note to Teacher:

The hydrogen atoms appear to have changed size from reactants to products. This is an artifact of the simulation from which the images come. You can challenge your students to think about whether their mass measurements of LEGO models in Lesson 2.2 support or refute this artifact. (Mass measurements suggest that atoms do not change mass during chemical reactions.)

2. Use Figure 2.2 as you discuss the following questions with your team and record your ideas.

   a. Why can only 10 H_2O molecules be made if there are still 5 O_2 molecules left?

   We run out of the hydrogen molecules. We use 2 hydrogen molecules for every 1 oxygen molecule, so we will use up 10 H_2 molecules and 5 O_2 molecules.

Table 2.7 below shows the number of atoms and molecules represented in Figure 2.2 before and after hydrogen molecules and oxygen molecules react to form water molecules. The number of atoms and molecules has been calculated for some of the cells in the table.

Table 2.7. Comparing Molecules and Atoms of Reactants and Products + Leftover Reactants

		Number of Molecules		Number of Atoms
Reactants	Hydrogen	10	Hydrogen	20
	Oxygen	10	Oxygen	20
	Water	0		
	Total number of molecules	20	Total number of atoms	40
Products + Leftover Reactants	Hydrogen	0	Hydrogen	20
	Oxygen	5	Oxygen	20
	Water	10		
	Total number of molecules	15	Total number of atoms	40

3. Complete Table 2.7 and use it as you discuss these questions with your team and record your ideas.

   a. How many *molecules in total* are there before the reaction? How does that compare with the number of molecules there are after the reaction?

   There are 20 molecules total before the reaction and 15 molecules total after the reaction.

   b. After the chemical reaction takes place, how many hydrogen atoms are present? How many oxygen atoms are there?

   20 hydrogen atoms and 20 oxygen atoms

   c. How does the number of atoms in the reactants of the chemical reaction compare with the number of atoms in the products?

   The number is the same—40 atoms of reactants and 40 atoms of products.

Teacher Talk and Actions
Activity 1 (cont)
Small group/partners: Students complete Steps 2 and 3.

Now that we've looked at the atoms and molecules, let's investigate the mass. Calculate the mass of the reactants and products in Table 2.8. An example has been done for you.

Table 2.8. Comparing Mass of Reactants and Products + Leftover Reactants

Atoms Making Up Reactants	# of Atoms	Mass of 1 Atom	Mass of Atoms in Reaction
Hydrogen	20	1 amu	20 amu
Oxygen	20	16 amu	320 amu
		Total mass of reactants	340 amu
Atoms Making Up Products + Leftover Reactants	**# of Atoms**	**Mass of 1 Atom**	**Mass of Atoms in Reaction**
Hydrogen	20		20 amu
Oxygen	20		320 amu
		Total mass of products	340 amu

4. Complete Table 2.8 and use it as you discuss these questions with your team and record your ideas.

   a. What do you notice about the total mass of the atoms before and after the reaction?

   The total mass of atoms (and mass of each kind of atom) is the same.

   b. What atomic mass did you use to determine the mass of all the hydrogen atoms in products + leftover reactants? Which science idea supports your decision?

   1 amu The idea that the mass of an atom doesn't change during a chemical reaction (Science Idea #8).

5. Answer this question on your own: Is total mass conserved during chemical reactions because atoms are conserved, because molecules are conserved, or because both atoms and molecules are conserved? Use information from Figure 2.2, Table 2.7, and Table 2.8 to support your answer.

   Total mass is conserved because atoms are conserved, meaning the total number of each type of atom stays the same. Regardless of how they are arranged in the reactants, products, and leftovers, the number of each type of atom that we start with (in reactant molecules) is the same as the number of each type of atom we end up with (in products and any leftover reactants). For example, the chart shows we start with 20 H atoms and 20 O atoms and end with 20 H atoms and 20 O atoms. The number of molecules is not necessarily conserved. In this case we start with 10 molecules of H_2 and 10 molecules of O_2 (20 molecules total) but end up with 10 molecules H_2O and 5 leftover O_2 molecules (15 molecules total).

Teacher Talk and Actions
Activity 1 (cont)
Individuals: Students answer the questions in Step 5 on their own. Whole class: Lead discussion. Invite student volunteers to share ideas about Step 5. Students should reach consensus that • the number of molecules is not conserved during chemical reactions and • the number of each type of atom is conserved, hence the total mass of those atoms, during chemical reactions.

Activity 2: Using Models to Think About Chemical Reactions in Opened Containers

Materials
For each team of students
1-gallon zipper plastic bag
LEGO model kit
Chemical Reaction Mats: Steel Wool (Iron) and Air (Oxygen), Baking Soda and Vinegar
Access to an electronic scale

In this activity, you will use models to help you think about why measured mass can change when a container is opened after a chemical reaction has taken place.

First, your teacher will demonstrate what happens when baking soda and vinegar are mixed in a sealed bag and then that bag is opened.

Remember, we have used two different ways to represent chemical equations, one using words and another using symbols. For example, the reaction between baking soda and vinegar can be represented by the following equations:

sodium bicarbonate (baking soda) + acetic acid (vinegar) ⟶ sodium acetate + water + carbon dioxide

$$NaHCO_3 + HC_2H_3O_2 \longrightarrow NaC_2H_3O_2 + H_2O + CO_2$$

Procedures and Questions

1. Watch as your teacher represents the chemical reaction between baking soda and vinegar with models. Record your data in Table 2.9.
   a. Some LEGO models of the molecules of the reactants are placed in a sealed container (the plastic bag, which represents the sealed bottle in Lesson 2.1).
      i. Count and record the number of models of molecules of the reactants in the bag.
      ii. Count and record the number of each kind of atom model in the bag.
      iii. Weigh and record the initial mass of the sealed bag containing the models.
   b. The chemical reaction is modeled in the sealed bag.
      i. Count and record the total number of models of molecules (reactants and products) in the bag.
      ii. Count and record the number of each kind of atom model in the bag.
      iii. Weigh and record the final mass of the sealed bag containing the models.

Teacher Talk and Actions
Activity 2 **(30 min)**
Whole class: Read introductory text aloud to the class. Demonstrate Step 1. If possible, use a document camera to project your demonstration. Otherwise, we suggest showing the Video—*Representing Baking Soda and Vinegar With LEGO Models in Sealed and Opened Container With Mass* to maximize visibility and, therefore, student engagement. Stop at appropriate times to allow students to make observations and record them in Table 2.9.

c. The bag is opened, and models of gas molecules (reactants or products) can enter or leave. In this case, some models of carbon dioxide molecules leave the bag.

   i. Count and record the number of models of molecules (reactants and products) of the substances in the bag.

   ii. Count and record the number of each kind of atom model in the bag.

   iii. Weigh and record the final mass of the opened bag containing the models.

Table 2.9. Baking Soda and Vinegar

	Number of Reactant and Product Models in the Bag	Number of Bricks in the Bag	Mass of the Bag (g)
a.	# $HC_2H_3O_2$ models: 4 # $NaHCO_3$ models: 5 # $NaC_2H_3O_2$ models: 0 # H_2O models: 0 # CO_2 models: 0	# H bricks: 21 # C bricks: 13 # O bricks: 23 # Na bricks: 5	Initial mass of sealed bag: 146.66/7 g
b.	# $HC_2H_3O_2$ models: 0 # $NaHCO_3$ models: 1 # $NaC_2H_3O_2$ models: 4 # H_2O models: 4 # CO_2 models: 4	# H bricks: 21 # C bricks: 13 # O bricks: 23 # Na bricks: 5	Final mass of sealed bag: 146.66/7 g
c.	# $HC_2H_3O_2$ models: 0 # $NaHCO_3$ models: 1 # $NaC_2H_3O_2$ models: 4 # H_2O models: 4 # CO_2 models: 0	# H bricks: 21 # C bricks: 9 # O bricks: 15 # Na bricks: 5	Final mass after bag is opened: 120.66/7 g

2. Discuss these questions with your team and record your ideas.

   a. Explain why the measured mass was constant while the bag was sealed.

   All of the atoms of reactant molecules and product molecules were trapped. No atoms could come in or go out.

   b. Explain why the measured mass decreased after the bag was opened. Where did the mass go?

   There were 4 carbon dioxide molecules that could leave the bag as a gas, so 8 oxygen atoms and 4 carbon atoms left.

Teacher Talk and Actions
Activity 2 (cont)
Whole class: Continue the demonstration or video, stopping at appropriate times to allow students to make observations and record them in Table 2.9. Small group: Allow students time to complete Step 2. Whole class: Lead discussion. Invite student volunteers to share ideas about Step 2. Students should reach consensus that carbon dioxide molecules (carbon and oxygen atoms) left the opened container, which caused the measured mass to decrease. Ask questions to further elicit and probe students' understanding of the relationship between measured mass and total mass: • When the container is sealed, how does measured mass compare with total mass? Why, in terms of atoms? • When the container is opened and carbon dioxide molecules leave, how does measured mass compare with total mass? Why, in terms of atoms?

Now your team will use LEGO bricks to represent the reaction between iron in steel wool and oxygen in air when steel wool is sealed in a flask.

The reaction between iron and oxygen gas can be represented by the following two models (equations):

$$\text{iron} + \text{oxygen} \longrightarrow \text{iron oxide (rust)}$$

$$4\,Fe + 3\,O_2 \longrightarrow 2\,Fe_2O_3$$

3. Build a model to represent the chemical reaction between iron and oxygen. Record your data in Table 2.10. Build models of seven O_2 molecules. Some of these will represent oxygen molecules in the air inside the sealed container (the bag, which represents the sealed flask). Some will represent the oxygen molecules in the air outside of the container.
   a. Seal some models of the reactant molecules in the container. Use eight Fe atoms and three O_2 molecules. Set aside the other O_2 molecules to represent oxygen molecules in the air outside of the container.
      i. Count and record the number of models of the reactant molecules in the bag.
      ii. Count and record the number of each kind of atom model in the bag.
      iii. Weigh and record the initial mass of the sealed bag containing the models.
   b. Model the reaction in the sealed bag—do not let any matter enter or leave the bag. Make sure that the models of reactants that react are completely used and the models of products are completely formed. That means there should be no parts of molecules.
      i. Count and record the number of models of molecules of the substances.
      ii. Count and record the number of each kind of atom model in the bag.
      iii. Weigh and record the final mass of the sealed bag containing the models.
   c. Open the bag and let the models of gas molecules (reactants or products) enter or leave. In this case, some models of oxygen molecules enter the bag.
      i. Add more O_2 molecules and react them with the Fe atoms. Keep adding O_2 until there are no unreacted Fe atoms left.
      ii. Count and record the number of models of molecules of the substances in the bag.
      iii. Count and record the number of each kind of atom model in the bag.
      iv. Weigh and record the final mass of the opened bag containing the models.

Teacher Talk and Actions
Activity 2 (cont)
Small group: Students complete Step 3. Provide as much scaffolding as you think is necessary to support partners in completing the activity successfully; keep all groups on pace. Remind students of proper use of the electronic scales (on/off, tare/zero, etc.) Small group: Students should stop at appropriate times to make observations and record them in Table 2.10.

Table 2.10. Steel Wool (Iron) and Air (Oxygen)

	Number of Reactant and Product Models in the Bag	Number of Bricks in the Bag	Mass of the Bag (g)
a.	# Fe bricks: 8 # O_2 models: 3 # Fe_2O_3 models: 0	# Fe bricks: 8 # O bricks: 6	Initial mass of sealed bag: 31 g
b.	# Fe bricks: 4 # O_2 models: 0 # Fe_2O_3 models: 2	# Fe bricks: 8 # O bricks: 6	Final mass of sealed bag: 31 g
c.	# Fe bricks: 0 # O_2 models: 0 # Fe_2O_3 models: 4	# Fe bricks: 8 # O bricks: 12	Final mass after bag is opened: 45 g

4. Discuss these questions with your team and record your ideas.

   a. Explain why the measured mass was constant while the bag was sealed.

   All of the atoms of reactant molecules and product molecules were trapped. No atoms could come in or go out.

   b. Explain why the measured mass increased after the bag was opened. Where did the additional mass come from?

   Oxygen atoms (from oxygen gas) could come into the bag. Two more oxygen molecules entered the bag. Since each oxygen molecule (O_2) is made up of 2 oxygen (O) atoms, that means 4 O atoms entered the bag. The four oxygen atoms that entered the bag reacted with leftover iron (Fe) atoms and formed two more iron oxide (Fe_2O_3) molecules.

5. Answer these questions individually.

   a. Based on what you observed when you modeled the chemical reaction between baking soda and vinegar and the chemical reaction between iron and oxygen, write a general statement that explains why the measured mass can change when a chemical reaction is carried out in an open system.

   The measured mass can change if substances enter or leave the system. If a gas produced in a chemical reaction can leave the system, the measured mass will decrease. If a gas can enter the system and react with substances already in the system, the measured mass will increase. When the gas enters and reacts, its atoms are now "trapped" in molecules of new substances that can't leave the system.

   b. Explain why a change in measured mass does not violate ideas about why mass is conserved.

   Atoms aren't created or destroyed during chemical reactions. An increase in measured mass is because more atoms are now part of the system being weighed on the balance. A decrease in measured mass is because atoms have left the system and are no longer being weighed on the balance.

Teacher Talk and Actions

Activity 2 (cont)

Small group:

Students complete Step 4.

Whole class:

Lead discussion. Invite students to share ideas about Step 4.

Step 4b. Explain why the measured mass increased after the bag was opened. Where did the additional mass come from?

Students should reach consensus that oxygen gas molecules (oxygen atoms) entered the opened container. They may neglect the idea that those very oxygen atoms reacted with leftover iron atoms to produce more iron oxide, which caused the measured mass to increase.

Ask questions to further probe students' understanding of the relationship between measured mass and total mass, as well as these questions:

• When the container is sealed, how does measured mass compare with total mass? Why in terms of atoms?

• When the container is opened and oxygen molecules enter, how does measured mass compare with total mass? Why in terms of atoms?

• Do oxygen atoms themselves cause the increase in mass we observe when steel wool rusts? What has to happen to the oxygen atoms for a change in mass to be detected?

Step 5a: To see if students are really thinking with an atomic/molecular model, ask what would happen if the new substance produced was also a gas and let small groups wrestle with it. Answer: If the new substance is also a gas, then it can leave the system. In such a case, whether the mass will increase depends on how heavy the molecules that enter are compared with the molecules that leave).

Science Ideas

The activities in this lesson are intended to help you construct an important idea about why measured mass can change. Read the idea below. Notice that Science Idea #10 explains observations about changes in mass in terms of atoms. We can observe when the mass changes, but we can't see atoms entering or leaving the system. However, because Science Idea #10 states a general principle that is consistent with a wide range of observations and data, we can use it to reason that since atoms are not created or destroyed in chemical reactions and the mass of an atom does not change, whenever we see the measured mass change we know that atoms have entered or left the system. You will be expected to use ideas about atoms to explain phenomena involving changes in measured mass.

Look back through Lesson 2.3 so far. In the space provided after the science idea, give at least one piece of evidence that supports the idea.

Science Idea #10: The *measured* mass of reactants and products is not always the same as the *total* mass. The measured mass changes if reactants or products (often gases) enter or leave the system. This is because atoms that make up reactants or products enter or leave the system.

Evidence:

The models showed us that when the container was sealed, the measured mass of the models was the same as the total mass of the models. This is because the total number of each type of atom was the same. When the container was opened and models of gas molecules could leave (like the carbon dioxide gas produced) or enter and react to form more product (like oxygen gas reacting with iron to form iron oxide), the total mass was not the same as the measured mass. Atoms were still conserved, but some were not included in the final measurement (like the carbon and oxygen atoms that make up the carbon dioxide gas that left the container) or in the initial measurement (like oxygen atoms that make up the oxygen gas that entered the container).

Teacher Talk and Actions
Science Ideas **(5 min)**
Whole class: Tell students: You have used models to help you understand important ideas about chemical reactions and mass. Read the science idea(s) aloud to the class. Add to the list of charted science ideas. Give students time to write down at least one example that supports the idea. Remind students that the science ideas will be used in constructing scientific explanations and arguments.

Activity 3: Explaining Changes in Mass

We can use the ideas we have learned in Lessons 2.1, 2.2, and 2.3 to explain how measured mass can change (or not change) during a chemical reaction.

Procedures and Questions

Some students worked together to answer the question below and explain their answer:

> Why did the measured mass increase when steel wool rusted if the total mass of reactants and products stayed the same (was conserved)?

1. Read the explanation, highlight the evidence, and underline the science ideas cited.

Explanation:

The measured mass increased because O atoms from oxygen gas (O_2) entered the container and reacted with Fe atoms from the steel wool to form rust (Fe_2O_3). The measured mass can increase during a chemical reaction if atoms enter the system and react to form molecules that can't leave (Science Idea #10). Atoms have mass, so the more atoms there are in the container, the more its measured mass will be.

Our models showed us that if more O atoms enter the container and react with Fe atoms, more Fe_2O_3 molecules can form and the measured mass can increase. So I think that's what happened with the real atoms. Once the O atoms that entered as oxygen gas were part of the rust, they were no longer a gas and therefore couldn't leave the container.

Evaluate the explanation according to the Explanation Quality Criteria.

Teacher Talk and Actions

Activity 3 (15 min)

Whole class: Read the introductory sentence aloud to the class.

Individuals: Students independently review the parts of the explanation.

Teacher Edition Discussion Notes:

This is a difficult but very important task. The explanation meets some but not all of the Explanation Quality Criteria.

1. The explanation does not provide any evidence. Carefully discuss how each piece of data (there are 3 pieces) provides evidence to support the explanation. During discussion, ask students what each highlighted piece of data is evidence of.

- The evidence supports the explanation: According to Science Idea #10, if atoms enter a system and react, then the measured mass of the system can increase. So the evidence must show that atoms have entered the container and reacted. The splint tests show that (a) oxygen was present in the closed container before the reaction and (b) oxygen was not in the closed container after the reaction. These 2 pieces of data provide evidence that oxygen gas was used up in the reaction, i.e., that oxygen gas reacted with the iron. The third piece of data, that more rust formed after the container was opened, is evidence that more iron and oxygen gas reacted. Since the oxygen in the closed container was used up, the additional oxygen must have come in when the container was opened.
- Students could point to the calculations they did in Lesson 2.2 for the baking soda and vinegar reaction and then reason that the opposite would happen for the iron rusting reaction because atoms enter the system.

2. Discuss the role that models play in the explanation.

- The models are consistent with the idea that measured mass increases if atoms enter a system and react. In Lesson 2.2, Pulling It Together Question 2 Table 2.6, we showed that the mass of the open container would decrease when baking soda reacts with vinegar because C and O atoms from CO_2 left the container. If we were to make a similar chart for the iron rusting reaction, we would have more atoms in the container after the reaction occurred. Therefore, the container would weigh more because O atoms from outside the container entered and reacted with Fe.

3. Carefully discuss whether <u>all</u> relevant science ideas are stated and if each science idea is just stated or is actually <u>used</u> to support the explanation.

- The explanation states the idea that measured mass can change during a chemical reaction if atoms enter or leave the system (Science Idea #10) AND uses the idea to show why the evidence supports the explanation: The first sentence applies the science idea to the iron rusting example.
- However, the explanation does not state or use the ideas that during a chemical reaction, atoms aren't created or destroyed (Science Idea #7) and the mass of an atom doesn't change (Science Idea #8), so the total mass of all the atoms doesn't change during a chemical reaction (Science Idea #9).

2. Use the explanation of why the measured mass increased when iron rusted to help you explain why the measured mass decreased when baking soda reacted with vinegar in an open container. Check to be sure your explanation meets the Explanation Quality Criteria.

Explanation:

The measured mass decreased when baking soda and vinegar reacted in an open container because the carbon dioxide gas that was produced in the reaction left the container, and the C and O atoms that make up these carbon dioxide molecules were no longer being weighed. My evidence that carbon dioxide gas was produced is that the gas gave a positive limewater test. My evidence that the carbon dioxide left the bottle is that we heard a whoosh sound when we opened the bottle and only then saw that the measured mass decreased. This is consistent with the idea that the measured mass can change if substances enter or leave a system, because atoms making up these substances enter or leave the system (Science Idea #10). Models showed how it was possible for the total mass to remain constant even though the measured mass decreased: The total number of C, H, O, and Na atoms at the beginning was the same as the total number of C, H, O, and Na atoms at the end, which is consistent with the idea that atoms are not created or destroyed in chemical reactions (Science Idea #7). And the mass of each brick stayed the same, so the total mass of all the bricks stayed the same. This is consistent with the ideas that the mass of an atom doesn't change (Science Idea #8), so the total mass of all the atoms doesn't change (Science Idea #9). However, at the end of the reaction some of the C and O bricks weren't being measured, because they had left the system as CO_2. This is consistent with idea stated above (Science Idea #10). So I think that's what happened to the real atoms.

3. Why doesn't an increase or a decrease in measured mass contradict the science idea that matter is conserved?

The total mass is conserved during a chemical reaction because the total number of each type of atom is conserved (Science Idea #9). This is because the total number of atoms doesn't change (Science Idea #7) and the mass of each atom doesn't change (Science Idea #8). The measured mass can change if atoms enter or leave the system (Science Idea #10). In other words, the atoms are always somewhere, but sometimes we just aren't weighing them on a balance.

Teacher Talk and Actions

Activity 3 (cont)

Here is a complete high-quality explanation for iron rusting:

> *The measured mass increased because more atoms were being measured. I think when the container was opened, more O atoms from oxygen gas (O_2) entered the container and reacted with Fe atoms from unreacted steel wool to form more rust (Fe_2O_3). Atoms aren't created during a chemical reaction (Science Idea #7) and the mass of an atom doesn't change during a chemical reaction (Science Idea #8), so the total mass doesn't change during a chemical reaction (Science Idea #9). If the measured mass increases, there must be more atoms in the system (Science Idea #10). Before the reaction, the air in the container tested positively for oxygen gas (splint test). After the reaction in the closed container, the splint test was negative and some rust had formed. This is evidence that the oxygen was used up to form the rust. After opening the container, we saw that more rust formed, so more of the chemical reaction occurred. Therefore, more oxygen gas must have come into the container to react with the remaining steel wool. Our models showed us that if more O atoms enter the container and react with Fe atoms, more Fe_2O_3 molecules can form and the measured mass can increase. O atoms weren't created and their mass didn't change, so the total mass of the O atoms didn't change, they just weren't measured until they were trapped in the additional Fe_2O_3.*

Partners: Students complete Step 2, writing an explanation for the decrease in measured mass observed when the container was opened after baking soda and vinegar chemically react.

Pulling It Together

Work on your own to answer these questions. Be prepared for a class discussion.

1. Now that you have modeled chemical reactions in opened containers, how would you answer the Key Question, **If atoms and total mass are always conserved during chemical reactions, why can measured mass change when the container is opened?** Explain it as you would to a friend who missed class today.

   Explain both cases—why measured mass sometimes increases and why it sometimes decreases when the container is opened.

   During chemical reactions, the products that form are made from the same atoms as reactants but are grouped differently. The mass of each atom and the total number of each atom does not change when they are rearranged, so the total mass does not change. The measured mass sometimes changes when the container is opened because reactants or products can enter or leave the container. This means that the atoms that make up those reactants or products (typically gases) enter or leave the container. If reactants enter the system and react to form more products, the number of atoms in the system increases and the measured mass increases. If products leave, the number of atoms in the system decreases and the measured mass decreases.

2. Think back to the video you saw in Chapter 1 of how nylon is made.

   a. If all of the liquid reactants in the beaker were used to make nylon thread and water, would the number of atoms in the nylon and water be less than, the same as, or more than the number of atoms originally in the beaker? Explain why.

   Before the nylon thread was pulled from the beaker, the number of atoms in the beaker would be the same. Unless atoms leave the beaker, the number of atoms won't change. Since no gases are involved in the reaction, no atoms should enter or leave the beaker. The products that form (nylon and water) are made from the same number of each type of atom (C, H, O, N) as the reactants but are arranged in different groups. The total number of each type of atom and the mass of atoms do not change when they are rearranged. So the number of atoms in the nylon and water are the same as the number of atoms originally in the beaker.

   b. As the nylon thread is pulled out of the liquids in the beaker, the measured mass of the liquids in the beaker decreases. Explain why in terms of atoms.

   When reactants or products can enter or leave the container when it is opened, then atoms can enter or leave the container. If products can leave (nylon is pulled out), atoms leave and measured mass decreases.

Teacher Talk and Actions

Pulling It Together **(15 min)**

Individuals:

Direct students to answer the Pulling It Together questions on their own, in detail.

Whole class:

Direct students to look back at their initial ideas about the Key Question and consider how their ideas have changed toward the Key Question/Pulling It Together Question 1.

Lead discussion. Invite students to share ideas about Pulling It Together Question 1 (the Key Question).

Students should reach consensus that if gas products (hence, atoms of products) leave an opened container, the measured mass will decrease. Students should reach consensus that if gas reactants (hence, atoms of reactants) entered the opened container, the measured mass will increase.

Challenge students as needed to ensure they understand that atoms that enter must react with other reactant atoms to make more products. If the atoms don't react, they can leave the container. When they react, they become "trapped" in different molecules that can't leave (unless one or more of the products is also a gas).

3. When a log burns in a fireplace, two gases (carbon dioxide and water vapor) are made. Imagine that you and a friend are talking about what happens when a log is burned in a fireplace. She thinks that the matter that makes up the log is destroyed. She thinks this because the log weighs much more than the ashes that are left over after the log burns.

   a. You and your class complete the following table to help decide whether your friend is right. In the second column, fill in science ideas, evidence, or models that support the statements in the first column.

   For each supporting item that you write, indicate whether it states the science idea, provides evidence, uses the science idea, or can be shown with a modeling activity. Hint: Some of the boxes will have more than one type of support for the statement.

Carbon dioxide and water vapor have different properties from the log.	*Evidence: Carbon dioxide and water vapor are both clear, colorless gases. The wood that makes up the log is a solid.*
Because carbon dioxide and water have different properties from the log, a chemical reaction must have occurred.	*Uses science idea: The ending substances of a chemical reaction can be recognized as new substances because they have different properties (Science Idea #2).*
Carbon dioxide and water vapor have different properties from the log because they are made of different arrangements of atoms.	*See opposite page for answer.*
In chemical reactions, atoms of the starting substances disconnect and reconnect in new ways to form ending substances.	*States science idea: During chemical reactions, atoms that make up molecules of starting substances disconnect from one another and connect in different ways to form molecules of ending substances ... (Science Idea #5).*
Atoms are not created or destroyed in chemical reactions.	*States science idea: Atoms are not created or destroyed during chemical reactions, so the total number of each type of atom remains the same ... (Science Idea #7).*
Because the atoms just rearranged, the total mass after the reaction is the same as the total mass before the reaction.	*See opposite page for answer.*

Teacher Talk and Actions

Pulling It Together (cont)

• While you might want to give credit if students are reasoning from models, press them to use the science ideas by asking, "Does any science idea support your statement?"

• Evidence for carbon dioxide and water vapor being gases could also cite evidence from the substance card that the boiling point of carbon dioxide is lower than room temperature, so it should be a gas. Students could also cite the boiling point of water and indicate that since a fireplace can boil water, a fire must be hotter than 100° C, so water would be a gas.

• Students should already know that molecules of a gas can spread out, but this should be made explicit.

• Students might also cite smoke or the smell of burning wood coming out of the fireplace as evidence for the fireplace being an open system.

Answers for table on page 224:

Answer for Row 3:

Uses science idea: Each substance is made up of a single type of atom or molecule. The properties of a substance are determined by the type, number, and arrangement of atoms that it is made up of. Because no two substances are made up of the same arrangement of atoms, no two substances have the same set of properties. (Science Idea #4)

Answer for Row 6:

States science idea: Because the mass of a particular atom doesn't change and because the number of each type of atom doesn't change, the total mass of the matter stays the same even though atoms are rearranged in chemical reactions ... (Science Idea #9). OR

Models: When we modeled the baking soda + vinegar reaction, we saw that C, H, O, and Na atoms just rearranged. However, if we removed bricks representing CO_2 molecules, the bricks that remained wouldn't weigh as much.

Statement that will help explain your answer	Science idea, evidence, or modeling activity that supports the statement
Carbon dioxide and water vapor (the ending substances) are gases that can leave open systems.	Evidence: The question states that carbon dioxide and water vapor are gases. Uses science idea: Molecules of gas move in all directions (so they can leave open systems).
The fireplace is an open system, so gases can leave the system.	Evidence: Matter can leave the fireplace (so the fireplace is an open system). Uses science idea: Molecules of gas move in all directions (so they can leave open systems).

b. Based on the table, what would you tell your friend?

My friend is incorrect. Burning a log is an example of a chemical reaction and atoms are conserved during chemical reactions. Changes during which starting substances interact to form new substances are called chemical reactions. The ending substances can be recognized as new substances because they have different properties from the starting substances (Science Idea #2). Carbon dioxide and water vapor have different properties (both clear, colorless gases) from the wood (a solid) so they are different substances, so a chemical reaction occurred. During chemical reactions, atoms making up molecules of starting substances rearrange to form molecules of ending substances (Science Idea #5). So the C, H, and O atoms that left as CO_2 and H_2O were part of the molecules of the starting substances. Because the mass of a particular atom doesn't change and because the number of each type of atom doesn't change, the total mass of the matter stays the same even though atoms are rearranged in chemical reaction (Science Idea #9). So the number of C, H, and O atoms that make up the product molecules CO_2 and H_2O is the same as the number of C, H, and O atoms that made up the reactant molecules (whatever molecules make up the log). The mass of the ashes is less than the mass of the wood from which it came because the gaseous products (carbon dioxide and oxygen) left the fireplace.

Teacher Talk and Actions
Pulling It Together (cont)
(Intentionally left blank)

4. The Statue of Liberty is made up of copper (Cu atoms). But the statue doesn't have the shiny, orange-brown appearance of copper. Instead, it is green. Why? After being exposed for many years to oxygen and carbon dioxide in the air, a thin layer of green copper carbonate ($CuCO_3$ molecules) formed on the copper statue. Answer the following questions, being sure to use ideas about atoms in your explanation and to write an explanation that meets the Explanation Quality Criteria.

   a. Is the change a chemical reaction? Explain.

Yes, the change is a chemical reaction. A chemical reaction has occurred when the ending substances have different properties from the starting substances (Science Idea #2). The copper carbonate (green solid) produced has different properties from the copper (orange shiny metal) and carbon dioxide and oxygen (colorless gases) reactants, so we have evidence that a new substance formed. New substances form during chemical reactions because atoms that make up the molecules of the reactants rearrange to form the molecules of the products (Science Idea #5). From the question, we know that copper (Cu) + carbon dioxide, oxygen, and water vapor in the atmosphere react to form copper carbonate ($CuCO_3$). Molecules of the product copper carbonate are made up of the same atoms (Cu, C, and O) as the Cu, CO_2, and O_2 but the atoms are arranged differently. If we built LEGO models of Cu, CO_2, and O_2, we should be able to rearrange the bricks to form $CuCO_3$. So that's probably what happened to the Statue of Liberty. (Optional: I don't know what role the H_2O plays in the chemical reaction, because there aren't any H atoms in the product $CuCO_3$).

   b. Do you think the Statue of Liberty now has less mass than, the same mass as, or more mass than when it was first made? Explain.

The Statue of Liberty is like an opened container. C and O atoms (from oxygen, water, and carbon dioxide molecules) can enter from the surroundings and react with the copper. The copper carbonate that forms on copper is like rust that forms on iron. C and O atoms (carbonate) attach to the Cu atoms on the surface of the statue. So the copper statue now—with a coating of copper carbonate—has more atoms and therefore more mass than the copper statue when it was first made.

Teacher Talk and Actions

Pulling It Together (cont)

Pulling It Together Question 4 can be used as an embedded assessment question:

- Poll students to see who thinks it is/is not an example of a chemical reaction.
- Ask for volunteers of each "side" to explain their thinking.

- Poll students to see who thinks the mass is more, less, or the same.
- Ask for volunteers of each "side" to support their answer.

Students should reach consensus that it is a chemical reaction and that the mass of the statue is greater now than when it was first made because atoms from outside the system react to form a product that coats the statue.

Closure and Link (2 min)

Whole class:

Summarize the lesson:

Today we learned that measured mass can sometimes change if reactants or products (often gases) enter or leave, because they are made up of atoms that enter or leave.

Gases that enter will only cause the mass to change if they undergo a chemical reaction and form new products or if they are otherwise trapped in the container (by closing the container again and forcing/pumping gas in).

Link to the next lesson:

The chemical reactions we have studied are in some ways similar to changes that occur in the bodies of plants and animals. Ideas about atom rearrangement, atom conservation, and mass conservation will also help you understand growth and repair in living things.

Toward High School Biology

Chapter 3

Teacher Edition

Chapter 3 Overview

This chapter focuses on explaining the growth and repair of plant body structures, such as leaves, wood, bark, fruit, and roots. The body structures of both plants and animals are made up largely of polymers. Plant structures are mainly made up of carbohydrates. Plants take in inorganic substances—carbon dioxide and water—and use these to produce glucose monomers through chemical reactions (photosynthesis). Once the glucose monomers are synthesized, the monomers react to form carbohydrate polymers (and water), and the carbohydrate polymers (especially cellulose) are used to build plant body structures. Just like in nonliving systems, atoms are rearranged but not created or destroyed during the chemical reactions involved in plant growth. Plant growth is consistent with the idea that matter is conserved during chemical reactions. As plants grow, the atoms that contribute to their increased mass come from molecules that plants take in from their environment: carbon dioxide molecules from the air and water molecules from the ground.

Lesson 3.1 presents students with several examples of plant growth and repair (corn plants growing, a giant pumpkin growing, kudzu vines growing, giant sequoia trees growing, and a tree healing from a pruning wound) and asks them to think about what new materials are being made. It shows that carbohydrates make up the largest percentage of the mass of most plant body structures (as opposed to protein or fat), thereby providing a rationale for focusing on carbohydrates to explain plant growth in subsequent lessons.

Lesson 3.2 introduces students to three major carbohydrates: cellulose, starch, and glucose. Students examine data on the different properties of these carbohydrates and observe the fibrous quality of celery versus the solubility of glucose in water, and they are expected to conclude that these carbohydrates are different substances. They are expected to reason that cellulose is probably a better substance than glucose for building plant body structures because it is fibrous and not soluble in water. Then they examine the molecules' structures to see that glucose could be the monomer that is used to build cellulose and starch polymers.

Lesson 3.3 focuses students on figuring out how plants get the glucose monomers they need to build the carbohydrates that make up their body structures. Students use data about the atomic composition of plants and the substances they take in to reason that carbon dioxide and water are probably the reactants that form glucose. By modeling the reaction, they can see that oxygen molecules are also produced and atoms are neither created nor destroyed. Students reason from data from isotopic labeling experiments to conclude that because CO_2 provides both the carbon and the oxygen atoms for glucose, carbon dioxide contributes most of the mass to glucose ($C_6H_{12}O_6$)—this confronts the misconception that gases are insubstantial and could not possibly contribute the mass to a giant tree.

Lesson 3.4 ties together the plant growth story line by providing experimental evidence (from isotopic labeling) and engaging students in a modeling activity to show that it is possible to make cellulose polymers from glucose so that the cellulose polymers are available for growth and repair of plant body structures.

Lesson 3.5 functions as an embedded assessment that requires students to synthesize ideas and practices from Chapters 1, 2, and 3. Students analyze and critique the scientist Van Helmont's explanation of the growth of a willow tree based on whether or not his evidence supports his conclusion. Then they write a better explanation and support it with evidence and reasoning from science ideas and models.

Chapter 3 Background Knowledge for Teachers

However complex the workings of living organisms, they operate according to the same principles of conservation and transformation of matter and energy as do all other natural systems. However, the chemical composition of living organisms is somewhat different from the physical environment in which they live. Most of the molecules making up the bodies of living organisms are molecules with carbon atoms linked together in chains or rings. In these molecules, carbon atoms are relatively reduced, meaning they are hydrogenated. Many of these molecules also contain nitrogen. In contrast, carbon and nitrogen in nonliving matter usually occur in oxidized form in simple molecules such as carbon dioxide (CO_2) and molecular nitrogen (N_2) in the atmosphere and carbonates (CO_3^{-2}) and nitrates (NO^{-1}) in Earth's crust.

Chapter 3 presents a basic story about the chemical reactions that contribute to the growth of plants. Plants take in carbon dioxide and water from their environment. Inside plants, these substances react to form glucose monomers and oxygen molecules. The glucose monomers can then react to form cellulose polymers (and water molecules), and these cellulose polymers make up the bulk of plant body structures.

Carbohydrates

Carbohydrates are the most abundant class of biological molecules. They are the major structural components of plants. The unit limits its treatment of carbohydrates to three: the polymers (polysaccharides) cellulose and starch, and the monomer (monosaccharide) glucose. It puts the most emphasis on cellulose as a contributor to plant body structures. Starch is typically used as a way to store glucose for later use (usually to burn via cellular respiration to release energy), while cellulose makes up the structures of plants. Because the unit does not focus on energy or storage, the function of starch is omitted. Lignin, a carbohydrate polymer that is a major component of wood, is omitted due to its complicated structure. Wood is made up of cellulose embedded in lignin. The growth of mushrooms is explored in Chapter 4 in a capstone lesson that incorporates ideas about growth from Chapters 3 and 4. Mushroom growth—that is, the growth of an organism that is neither plant nor animal—involves the production of chitin, a carbohydrate polymer that also contains nitrogen atoms. The structure of chitin is omitted to avoid unnecessary detail.

Photosynthesis

Photosynthesis is a light-powered series of reactions carried out by plants, algae, and some bacteria. Starting with carbon dioxide and water, photosynthesis produces glucose monomers (which are retained in the organism) and oxygen gas (which is released into the environment). The unit provides a highly simplified story about photosynthesis. Students only encounter the net reaction:

$$6CO_2 + 6H_2O \rightarrow C_6H_{12}O_6 + 6O_2$$

However, photosynthesis is a series of reactions. When students encounter the isotopic labeling data in Lesson 3.3 (see Tables 3.6 and 3.7), there is a seeming inconsistency that may lead students to conclude, incorrectly, that some of the oxygen atoms making up O_2 come from H_2O.

Table 3.6 Location of Labeled Atoms (highlighted) at the Beginning and the End of the Experiment

	Starting Substances (Reactants)		**Ending Substances (Products)**	
Experiment 1	CO_2	H_2O	$C_6H_{12}O_6$	O_2

Table 3.7 Location of Labeled Atoms (highlighted) at the Beginning and the End of the Experiment

	Starting Substances (Reactants)		Ending Substances (Products)	
Experiment 2	CO_2	$H_2\mathbf{O}$	$C_6H_{12}O_6$	$\mathbf{O}_2$

The net equation has six O atoms (from six H_2O) but needs to make 12 O atoms (in six O_2 molecules). This is because the net reaction is the sum of two reactions: the light reaction requires an additional six H_2O (12 H_2O per six O_2), and the dark reactions produce six H_2O per glucose.

Thus,

$$6CO_2 + 126H_2O \rightarrow C_6H_{12}O_6 + 6O_2 + 6H_2O$$

Most students will not notice this, but a few might ask.

Because the unit does not deal with energy, the role of sunlight is not elaborated. Students focus on atoms, and they are told that sunlight is not made up of atoms and are asked to reason that it does not contribute to the mass of the products. Students are not asked to consider any photosynthesizing organisms that are not plants or algae.

Proteins

Although not covered in the unit, plants can take the glucose they make and a source of nitrogen (from minerals in the soil) and synthesize amino acids. These amino acids can react to form proteins, just as in animals. Although most plant body structures are made up of more carbohydrates than proteins, some structures, especially seeds (including nuts and beans), can contain a great deal of protein. Even though their bodies are made up mostly of proteins, herbivorous animals and vegan humans can live, grow, and repair on a plant-only diet because of the protein in plants.

Energy

Carbohydrates are both building blocks for complex molecules and also fuels to satisfy the energy needs of cells. The unit does not discuss the metabolism of carbohydrates to release energy for the cell. We made this decision for both practical and conceptual reasons. The time that school districts were able to devote to chemical reactions was insufficient to develop coherent stories of both matter transformation and energy transformation. Our observations indicated that students starting the unit lacked mental models of both matter transformation and energy transformation, and six weeks was not enough time to help students develop and use both. Because matter transformation could be made concrete with physical models, we opted to start there.

It is important to note that we initially tried to include a partial energy story—the rule that "atoms do not turn into energy, and energy does not turn into atoms"—but found it to be unsatisfying to most students. While a few students were able to cite the rule when challenging classmates' explanations that involved creating matter from energy, the rule didn't appear to convince their classmates. We think that students will be less likely to invoke matter-from-energy explanations of chemical reactions if they have a strong mental model of atom rearrangement and conservation.

Chapters 3 and 4 Commonly Held Student Ideas and Misconceptions

Researchers have documented several commonly held student ideas and misconceptions associated with plant and animal growth. The following is a brief summary of the incorrect ideas relevant to the unit's treatment of plant and animal growth.

Some incorrect ideas about plant and animal growth are revealed when students are asked questions at the macroscopic level (e.g., where the mass comes from when a plant or animal grows, and the relationship between food and growth). Other incorrect ideas are revealed when students are asked questions at the atomic/molecular level (e.g., whether plants and animals are made up of molecules, and the role of chemical reactions in growth).

Most of the initial studies probed students' ideas at the macroscopic level. When students' ideas about plant growth are probed at the substance level, responses indicate that some students think that soil is food for plants (Leach et al. 1992; Leach, Driver, Scott, and Wood-Robinson 1996; Vaz, Carola, and Neto 1997; Wandersee 1983), or that substances in the soil are food for plants (Kuech, Zogg, Zeeman, and Johnson 2003; Leach et al. 1992; Simpson and Arnold 1982; Stavy et al. 1987; Tamir 1989; Wandersee 1983). This may result from students' viewing food as anything an organism takes in, rather than as a substance that is a source of both matter and energy.

Middle school and high school students have difficulty thinking of the human body as a chemical system and have little knowledge about the elements composing the living body (Stavy, Eisen, and Yakobi 1987). In particular, middle school students think organisms and materials in the environment are very different types of matter and are not transformable into each other. For example, animals are made of bone, muscle, and skin; plants are made of leaves, stems, and roots; and the nonliving environment is made of water, soil, and air (Smith and Anderson 1986). Students holding such views might think that matter from food is destroyed and matter making up the body is created (e.g., from energy). Alternatively, some students "view food as being simply converted into ... waste materials and eliminated from the body" (Smith and Anderson 1986, p. 11).

Research has also shown that few students are able to apply ideas about atom rearrangement and conservation during chemical reactions to plant growth. Findings from Project 2061's assessment study (AAAS Project 2061, n.d.) revealed that students commonly

- do not associate growth with an increase in mass (and hence have no need to account for it);
- think that the mass of gases is inconsequential (and hence can't contribute significantly to growth); and
- do not associate the growth and functioning of organisms with chemical reactions.

All of these findings are consistent with students lacking a particulate model of chemical reactions and the ability to use it to think about chemical reactions associated with animal and plant growth. Chapters 1 and 2 have been carefully designed to help students develop a model of atom rearrangement and conservation. Chapters 3 and 4 support students as they learn to apply the model to growth and repair in living things.

References

AAAS Project 2061. n.d. AAAS science assessment. *http://assessment.aaas.*

Kuech, R., G. Zogg, S. Zeeman, and M. Johnson. 2003. Technology rich biology labs: Effects of misconceptions. Paper presented at the Annual Conference of the National Association for Research in Science Teaching (NARST), Philadelphia.

Leach, J., R. Driver, P. Scott, and C. Wood-Robinson. 1992. Progression in understanding of ecological concepts by pupils aged 5–16. Report, Children's Learning in Science Research Group, Centre for Studies in Science and Mathematics Education, University of Leeds. *www.stem.org.uk/elibrary/resource/26951.*

Leach, J., R. Driver, P. Scott, and C. Wood-Robinson. 1996. Children's ideas about ecology 2: Ideas found in children aged 5–16 about the cycling of matter. *International Journal of Science Education* 18 (1): 19–34.

Simpson, M., and B. Arnold. 1982. The inappropriate use of subsumers in biology learning. *European Journal of Science Education* 4 (2): 173–182.

Smith, E. L., and C. W. Anderson. 1986. Alternative student conceptions of matter cycling in ecosystems. Paper presented at the Annual Conference of the National Association of Research in Science Teaching (NARST), San Francisco, CA.

Stavy, R., Y. Eisen, and D. Yaakobi. 1987. How students aged 13–15 understand photosynthesis. *International Journal of Science Education* 9 (1): 105–115.

Tamir, P. 1989. Some issues related to the use of justifications to multiple-choice answers. *Journal of Biology Education* 23 (4): 285–292.

Lesson Guide
Chapter 3, Lesson 3.1
The "Stuff" That Makes Up Plants

Focus: This lesson focuses on the composition of plant bodies, emphasizing that they are made up of matter (atoms and molecules) and that carbohydrate molecules make up most of the mass of plant structures.

Key Question: What substances do plants need in order to grow and repair?

Target Idea(s) Addressed

- Plants build many different body structures as they grow and repair.
- These body structures are made up largely of carbohydrate molecules. Because they are different substances, they have different molecular structures and their glucose monomers are arranged in different ways.

Science Practice(s) Addressed

- Analyze and interpret data to provide evidence for phenomena.
- Apply scientific ideas, principles, or evidence to construct an explanation for real-world phenomena.

Materials

Activity 1: Projector, Video: *Corn Growth*, Slide show: *Plant Growth and Repair*

Activity 2: Slide show: *Plant Body Structures*

Advance Preparation

Locate the video and slides on the THSB website and set up the projector.

Neither the video nor slide shows will have audio, so review them and use the teacher facilitation pages to prepare to narrate them (or point out key parts) for students.

Organize materials in advance so that each set of partners or table teams has materials when class begins.

Phenomena, Data, or Models	Intended Observations	Purpose	Rationale or Notes
Activity 1 Video and photos of plants growing and repairing their bodies: corn plants, pumpkin plants, kudzu vines, giant sequoia trees, and a tree healing after a branch has been pruned	Several plants (corn, pumpkins, kudzu, and trees) grow and repair. These plants build new materials (stems, leaves, bark, fruit) when they grow and repair.	Students observe data to ensure that they are thinking of plants as having bodies, being made up of matter, and growing and repairing.	Corn plants and kudzu illustrate stem and leaf growth, pumpkin plant shows fruit growth, and sequoia trees emphasize huge mass increases accompanying growth. The healing tree shows new growth that covers over the wound.
Activity 2 Slide show of plant body structures	Plants have a variety of body structures made from a variety of materials.	Students become familiar with the fact that plants have complex bodies with specialized structures and that those structures need to be built as plants grow.	This might be obvious to some students, but not all will be as familiar with plant body structures as they are with animal body structures.
Activity 2 Data on protein, fat, and carbohydrate content for a variety of plants (broken down by structure)	Aside from water and minerals, plants are made up mainly of protein, fat, and carbohydrate molecules. Of these, carbohydrates are present in the highest amount.	Students use data to determine that carbohydrates are the most important contributor to plant growth. (Carbohydrates will remain the focus for the rest of Chapter 3.)	Knowing that plants are made up of a lot of carbohydrate molecules will drive the coming questions: What are carbohydrates and how do plants get the carbohydrates they need to grow?

Lesson 3.1—The "Stuff" That Makes Up Plants

What do we know and what are we trying to find out?

In the previous two chapters, you observed and modeled chemical reactions in nonliving systems because they were simpler systems than the bodies of plants and animals. By observing and modeling chemical reactions in simpler systems, you were able to figure out what happens to atoms and molecules during chemical reactions and to the measured mass of the system. You are now ready to apply what you learned about chemical reactions to the growth of plants. In order to think about how plants might grow, you need to know what plant body structures are made of. That will give you clues to what substances plants will need in order to grow.

Answer the Key Question to the best of your knowledge. Be prepared to share your ideas with the class.

Key Question: What substances do plants need in order to grow and repair?

Plants have a variety of parts/structures like stems, leaves, and roots. To grow, plants need to make more of these structures. They also need to make stuff like flowers and seeds. All this stuff is matter, so plants need more atoms from their environment.

Teacher Talk and Actions

What do we know and what are we trying to find out? (3 min)

Whole class:

Read text under "What do we know and what are we trying to find out?" aloud to the class.

Key Question (5 min)

Individuals:

Students respond to the Key Question: **What substances do plants need in order to grow and repair?**

Whole class:

Lead discussion. Invite student volunteers to share ideas about the Key Question.

The goal is primarily eliciting and probing student ideas here.

When students share their answers, do not indicate if their answers are right or wrong. However, do not let students assume by your silence that they were correct. Make it clear that students are expressing ideas, but those ideas may change based on what they will see in the lessons. (Remind students of this regularly.)

When students volunteer answers, ask "How do you know?" to find out whether they are using any ideas they have learned about chemical reactions (e.g., plants are matter, so they must be made of atoms; when plants grow they need to take in atoms from their surroundings to build body structures like leaves; to make new body structures they probably need to carry out chemical reactions) or if they are just responding in terms of what plants need to live and grow (e.g., if plants don't get water they will die).

If students volunteer that plants need energy or sunlight, ask whether sunlight is made up of atoms. If students know that sunlight is not made up of atoms, ask where the atoms plants need to build their body structures could come from. Students will probably know that atoms can come from water and from soil.

If students don't volunteer that air could be a source of atoms for building body parts, ask where the atoms for making rust came from. Note whether any students know that air can be a source of atoms.

Activity 1: Observing Plant Growth

To help us think about how plants grow, we'll start by observing examples of plant growth. As you watch the video and slides, think about the body structures being made as the plant grows or repairs itself after an injury.

Procedures and Questions

1. Observe the video and photos of plant growth. As you answer the questions below, think about the body structures plants have that you can see and also the body structures you can't see.

   a. What body structures are getting bigger as the **corn plants** grow?

   Stems, leaves, and roots

   b. What body structures are getting bigger as the **pumpkin** grows?

   Stems, leaves, pumpkin/fruit, seeds, and roots

   c. What body structures are getting bigger as the **kudzu vines** grow?

   Stems, leaves, and roots

   d. What body structures are getting bigger as the **giant sequoia trees** grow?

   Leaves (needles), stem/trunk, roots, wood, and bark

   e. What body structures are being rebuilt as the **tree heals a wound** after one of its branches was pruned (chopped off)?

   Wood and bark, stem/trunk

Teacher Talk and Actions
Activity 1 **(10 min)**
Read the text underneath Activity 1 aloud to the class. Play Video—*Corn Growth* for students. (Feel free to pause or reshow parts of it.) Have students write down as many structures as they can think of that are getting bigger when the corn grows. Show the slide show *Plant Growth and Repair* for students. Share the following contextual information with students: Pumpkins are fruits (they contain seeds). Certain varieties can grow extremely large. Kudzu is a vine native to Japan that is invasive in North America. It grows very quickly and can completely cover surrounding trees and shrubs. Giant sequoia trees are the world's largest individual trees by volume. (Photo is of two individual trees.) But they start as tiny seedlings like any tree. Trees usually "wall off" injured areas by building over them so that rot and decay cannot spread through the rest of the tree.

2. You have just observed that various plant body structures get bigger as plants grow, but how do we know that plants are actually adding new matter to their bodies? To answer this question, we will observe what happens to the mass of growing plants. Examine the data in Table 3.1 and answer the questions that follow.

Table 3.1. Typical Change in Mass (by Weight in Grams) of Some Growing Plants, Over Different Time Periods

Plant	Initial Weight (Age)	Final Weight (Age)	Change in Weight	Increase or Decrease?
Mung plants	24.6 g (1 week)	28.1 g (3 weeks)	3.5 g	+
Bean plants	47 g (3 weeks)	262 g (10 weeks)	215 g	+
Corn plants	0.7 g (1 week)	896 g (13 weeks)	895.3 g	+
Grass	2 g (60 days)	13 g (140 days)	11 g	+

a. Using the data in Table 3.1, calculate the change in weight for each plant (you can use a calculator). Put a plus sign (+) next to any change in mass that was an increase. Put a minus sign (–) next to any change in mass that was a decrease.

b. Based on these data, what happens over time to the mass of a plant when it grows?

It increases. Each of the four types of plants increased in weight, which means each gained mass.

Teacher Talk and Actions

Activity 1 (cont) (10 min)

Have students read Table 3.1 and let them use calculators to determine the change in weight for each plant, then mark a + or – to indicate an increase or decrease.

Note: Each of the plants was tracked over a different time span (due to different life spans and growth rates), so focus students on "initial" and "final" if they are getting confused by the times in parentheses.

Make sure that all students understand that plants increase in mass when they grow before you move on from this activity.

Give students time to answer the question in Step 2b.

Some students might be in the habit of saying, "It stays the same," when asked questions about mass or the number of atoms.

They are asserting an understanding of conservation of mass, but make sure they are paying attention to the question being asked. When an plant grows and gains weight, the measured mass of the plant increases.

Push students to think about the implications of an increase in weight:

If each of the four types of plants increased in weight, what is happening to the amount of matter as they grow? What must be happening to the number of atoms making up the plants' body structures? Where do you think these atoms could have come from?

Activity 2: The Substances That Make Up Plants

The videos and slides you observed in Activity 1 showed that plants produce a lot of different body structures as they grow. What substances are these body structures made up of? When you eat fruit like strawberries or tomatoes, or stems like celery or bamboo shoots, you have probably observed that lots of liquid comes out of them. Plant body structures are made up of a lot of water. But water doesn't give tomatoes or celery their texture, flavor, or nutritional value. If you have ever eaten packaged vegetables, you may have noticed the label on the package. The label indicates the amounts of the main nutrients—carbohydrates, fats, and proteins—plus a few other substances like sodium, cholesterol, and vitamins in a serving of the food. In this activity, you will find out how much of each of the main nutrients make up some common plant body structures.

Nutrition Facts	
Serving Size 1 cup raw green beans (100g)	
Servings Per Container 1	
Amount Per Serving	
Calories 30	Calories from Fat 0
	% Daily Value*
Total Fat 0g	**0%**
Saturated Fat 0g	**0%**
Trans Fat 0g	
Cholesterol 0mg	**0%**
Sodium 5mg	**0%**
Total Carbohydrate 7g	**2%**
Dietary Fiber 3g	**12%**
Sugars 3g	
Protein 2g	

Procedures and Questions

1. Observe the slides your teacher will show you of plant body structures.

2. Examine Table 3.2 with your group and answer the questions that follow.

Table 3.2. Relative Mass (in Grams per 100 Grams) of Carbohydrate Molecules, Fat Molecules, and Protein Molecules in Some Plant Body Structures

Structure	Plant	Carbohydrates (per 100 g)	Fats (per 100 g)	Proteins (per 100 g)
Root	Carrot	8.24	0.13	0.64
	Potato	17.00	0.10	2.00
Stem	Bamboo shoot	5.20	0.30	2.60
	Celery	2.97	0.17	0.69
Leaf	Spinach	3.63	0.39	2.86
	Lettuce	2.87	0.15	1.36
Flower	Broccoli	5.24	0.35	2.98
	Cauliflower	4.97	0.28	1.92
Fruit	Strawberry	7.68	0.30	0.67
	Tomato	3.89	0.20	0.88
Seed	Snap beans	6.97	0.22	1.83
	Kidney beans	4.10	0.50	4.20

a. In Table 3.2, **circle** the type of molecule that is present in the largest amount in most of the plants' body structures.

Teacher Talk and Actions	
Activity 2	**(20 min)**

Read the text underneath Activity 2 aloud to the class.

Project the slide show *Plant Body Structures* and quickly flip through it to refresh students on typical plant structures, if needed.

Small groups:

Give students time to read Table 3.2 and complete Step 2.

b. Are there any plant structures for which the molecule you circled is not present in the largest amount? If so, which one(s)?

Yes, kidney beans, a kind of seed

c. Based on the data in Table 3.2, what type of molecule do you think plants will need the most of in order to make their bodies? Why do you think so?

Carbohydrates, because almost all the plant body structures had more carbohydrate than protein or fat molecules

Teacher Talk and Actions	
Activity 2	**(20 min)**

Question 2b is not crucial for advancing the plant story line, but it may be useful in the animal chapter for explaining why people who are strict vegetarians (eat no animals or animal products) can still make proteins.

Lead a brief class discussion by asking some students to share their responses to the questions in Step 2.

If time permits, ask students if they have any ideas about what atoms make up carbohydrates and why they think so. Some students may realize that "carbo" comes from carbon atoms and hydrate comes from water, so carbohydrates are probably made up of C, H, and O atoms.

Pulling It Together

Work on your own to answer these questions. Be prepared for a class discussion.

1. Now that you have seen plants grow and learned about the substances that make up their body structures, how would you answer the Key Question, **What substances do plants need in order to grow and repair?**

Plants need mostly carbohydrates (and also some proteins and fats) to build their body structures. We've seen that plant body structures are made up mostly of carbohydrates, so if we want to figure out what plants need to build body structures, we should focus on carbohydrates.

2. Where do you think the atoms for plant growth come from? List all the places you think plants might get atoms from.

Plants could get atoms from the soil and from water. And just like we saw with iron rusting, plants could also get atoms from the air.

Teacher Talk and Actions

Pulling It Together (7 min)

Individuals:

Students answer Pulling It Together questions on their own, in detail.

Whole class:

Lead a class discussion.

Invite students to share ideas about Pulling It Together Question 1, which returns to the lesson Key Question. Direct students to look back at their initial ideas about the Key Question and consider how their ideas have changed.

Pulling It Together Question 2 elicits student ideas about what plants take in from their environment that might contribute to growth. Fully correct answers are not expected yet.

Closure and Link (3 min)

Whole class:

Summarize the lesson in a few statements:

We saw that plants build a variety of body structures as they grow and that those body structures—leaves, stems, roots, and flowers—are mainly made up of carbohydrate molecules.

Link to the next lesson:

In the next two lessons, we will investigate how plants get the carbohydrates they need to build their body structures.

Lesson Guide
Chapter 3, Lesson 3.2
Carbohydrates That Make Up Plants

Focus: This lesson introduces students to carbohydrate molecules: some of their functions, the basic molecular structures, the idea that there are different carbohydrate molecules, and that these carbohydrates are different substances.

Key Question: Are the carbohydrates that make up different plant structures the same substance?

Target Idea(s) Addressed

- Carbohydrate polymers are made of glucose monomers.
- Different carbohydrates have different arrangements of glucose monomers and, therefore, are different substances with different characteristic properties.

Science Practice(s) Addressed

- Analyze and interpret data to provide evidence for phenomena.
- Use a model to describe phenomena.
- Apply scientific ideas, principles, or evidence to construct an explanation for real-world phenomena.

Materials

Activity 1: Celery (1 stalk per group of 4 students), Glucose (1 teaspoon per class period), Clear container, Tap water

Advance Preparation

Organize materials in advance so that each set of partners or table teams has materials when class begins.

Phenomena, Data, or Models	Intended Observations	Purpose	Rationale or Notes
Activity 1 Data on the properties (solubility and ability to form fibers) of starch, cellulose, and glucose	Different carbohydrates have different properties and are, therefore, different substances.	Students observe that different carbohydrates are different substances and reason that some are better for building plant bodies than others.	Knowing that glucose is a different substance from cellulose will prepare students to expect that a chemical reaction occurs when glucose reacts to form cellulose (and water molecules).
Pulling apart a celery stalk	Celery is tough and fibrous.	After reading about fibers in a table, students experience an activity to help them understand what fibers are. (Students should reason that celery is mostly cellulose.)	Celery is a concrete visual and tactile way for students to experience what it means that cellulose "forms fibers."
Demo of glucose dissolving in water	Glucose is soluble in water.	Students observe first-hand evidence that glucose is soluble in water and reason that it is probably not useful for building plant body structures because they contain a lot of water.	Glucose dissolving uses a property students are familiar with from earlier chapters to visually illustrate why plant body structures are not mostly glucose. This motivates their investigation of how plants build cellulose from glucose.
Activity 2 Molecular structure of two carbohydrate polymers (starch and cellulose) and the monomer, glucose, they are made of	Carbohydrates are made from C, H, and O atoms. Different carbohydrate polymers can be made from glucose monomers.	Students use models to visualize carbohydrate polymers and the glucose monomers that make up the polymers.	Knowing the molecular structure of carbohydrate polymers will help students reason about what chemical reactions might occur in plants to build their body structures.

Lesson 3.2—Carbohydrates That Make Up Plants

What do we know and what are we trying to find out?

We saw in the last lesson that the different body structures of plants are mostly made up of carbohydrates, or "carbs," as they're often called. So to understand how plants grow, we'll need to take a closer look at the carbohydrates that make up plants. Think about the last time you bit into a ripe fruit, such as a tomato or strawberry. Did it taste sweet? Now compare the taste of the ripe tomato or strawberry with the taste of a stem of celery or a root vegetable, such as a potato. Were they as sweet? Next, compare what it feels like when you bite into a strawberry with what it feels like when you bite into a piece of celery. What do you think it would feel like if you tried to bite into the bark of a tree? All of these different plant structures are mostly carbohydrates, yet they are not all the same. In this lesson, we will examine ways that carbohydrates are similar to one another and how they differ.

Answer the Key Question to the best of your current knowledge. Be prepared to share your ideas with the class.

Key Question: Are the carbohydrates that make up different plant structures the same substance?

Maybe the difference in sweetness between a tomato and a piece of celery is because they are made up of different carbohydrates. However, celery and tomatoes are also partly made up of fats and proteins, so differences could be due to those substances. The only way to know whether the carbohydrates that make up stems and leaves are the same or different is to look at the properties of the carbohydrates that make up different plant structures.

Teacher Talk and Actions
What do we know and what are we trying to find out? (3 min)
Whole class: Read the text under "What do we know and what are we trying to find out?" aloud to the class. Terminology Note: Unlike the term *proteins,* which refers only to the polymer, the term *carbohydrates* includes both monomers and polymers. When referring specifically to polymers, *carbohydrate polymers* must be used.
Key Question (5 min)
Individuals: Students respond to the Key Question: **Are the carbohydrates that make up different plant structures the same substance?** Whole class: Lead discussion. Invite student volunteers to share ideas about the Key Question. The goal is primarily eliciting and probing student ideas here. There is no way to answer the question without knowing something about either the properties or the molecular composition of the carbohydrates that make up different plant structures, which students will observe in this lesson. It's important for students to realize that they would need more evidence to draw conclusions. Don't press students to draw conclusions without evidence.

Activity 1: Properties of Familiar Carbohydrates

Materials
For each team of students
1 celery stalk

You learned in Chapter 1 that the way to tell whether two substances were the same or different was to look at their properties. In this activity, you will examine the properties of three carbohydrates to decide whether any two are the same substance and consider what plants might use them for.

Procedures and Questions

1. Examine Table 3.3, and then work with your group to answer the questions that follow.

Table 3.3. Properties of Three Common Carbohydrate Molecules

Carbohydrate Molecule	Solubility in Water	Does It Form Fibers?
Glucose	High	No
Starch	Low	No
Cellulose	Low	Yes

a. Based on the data in Table 3.3, is glucose a different substance from cellulose? Why or why not?

Yes, it is, because glucose is soluble in water and does not form fibers, while cellulose is not soluble in water and does form fibers. Because they have these different properties, they must be different substances.

b. Is cellulose a different substance from starch? Why or why not?

Yes, it is, because cellulose forms fibers while starch does not. The difference in this property means they must be different substances.

2. Your teacher will hand out a celery stalk. Examine it with your group. Try pulling it apart in different directions or peeling off sections of it. The strands that come off are fibers.

a. Which of the carbohydrate molecules in Table 3.3 do you think gives celery its stringiness? Why do you think so?

Cellulose does, because it is the only one that forms fibers and the celery has a lot of fibers.

Teacher Talk and Actions	
Activity 1	**(12 min)**

Read the text underneath Activity 1 aloud to the class.

Small groups: Give students time to read Table 3.3 and complete Step 1.

Step 1b: Cellulose and starch differ by only one property (of the two properties students have data for). Because only one property differs, students may not be convinced that the molecules are indeed different substances. Remind students that any difference—even only one—in characteristic properties indicates a different substance.

Pass out one fresh celery stalk to each group of students. Instruct students to try pulling it apart in different ways, including peeling along the "grain" to separate strands. Tell students these strands are fibers.

b. Do you think that molecules that form fibers or those that do not form fibers would be better for building some plant structures? Why do you think so?

Molecules that form fibers might be better for building plant structures like stems and leaves because they are strong and tough to rip apart, as we noticed when we tried to rip apart the celery. However, plant structures like the roots of potato plants or the fleshy part of tomatoes and strawberries might not need to be so tough. These could be made mostly of a substance that doesn't form fibers, like starch.

3. Observe your teacher demonstrate adding pure glucose to water.

a. Record your observations of what happens below. Do your observations match the data in the table?

The sugar was poured into the water. At first you could see it, but as the teacher stirred the water, it all seemed to disappear, which meant it had dissolved. This matches the table, which says that glucose has high solubility in water.

b. Would glucose be good for building plant body structures? Why or why not?

No, it wouldn't, because plant bodies contain a lot of water, and glucose would just dissolve in all that water. However, plant parts like fruits that rely on animals to eat and disperse its seeds might have more glucose to persuade animals to eat it.

Teacher Talk and Actions
Activity 1 (cont)
Small groups: Give students time to complete Step 2. Step 3: Fill a clear container with tap water. Have the class watch while you pour a teaspoon of pure glucose into the water. Stir it around until it all dissolves. Encourage students to write descriptive observations. Give them time to complete Step 3.

Activity 2: Molecular Composition and Structure of Carbohydrates

You learned in Chapter 1 that if two substances have different properties, then they are made up of different molecules. And you learned that molecules can differ in the types of atoms that make them up. In this activity, you will have a chance to examine models of the molecules making up the three carbohydrates you just decided were different substances—glucose, starch, and cellulose.

Procedures and Questions

1. Examine the data in Table 3.4. What does it tell you about the types of atoms making up the three carbohydrates?

Glucose, starch, and cellulose are all made up of C, H, and O atoms in the ratio of about 1:2:1 (25:50:25 reduced to simplest whole numbers).

Table 3.4. Composition of Common Carbohydrates

Carbohydrate	% C Atoms	% H Atoms	% O Atoms	% Other Atoms	Total
Glucose	25.0	50.0	25.0	0	100
Starch	28.6	47.6	23.8	0	100
Cellulose	28.6	47.6	23.8	0	100

a. If glucose, starch, and cellulose are all made up of the same types of atoms, how might they differ?

Maybe the C, H, and O atoms are arranged differently in the three substances.

2. In Figures 3.1, 3.2 (p. 112), and 3.3 (p. 112), color the O atoms red. With your group members, carefully examine the figures and answer the questions that follow.

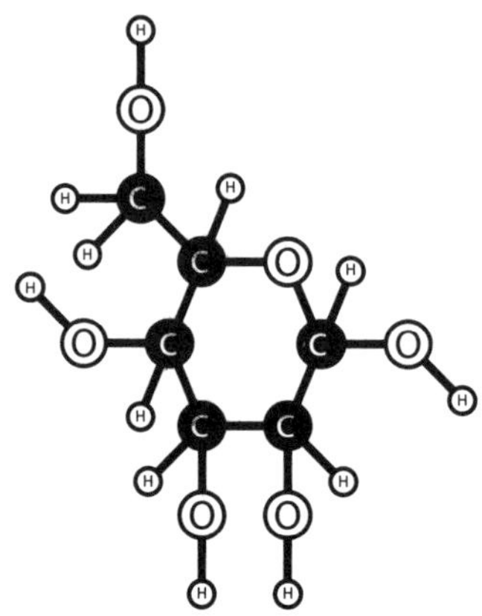

Figure 3.1. Model of the Structure of a Carbohydrate Molecule Called ***Glucose***

Teacher Talk and Actions
Activity 2 **(12 min)**
Whole class: Consider pointing out that the 2:1 ratio of hydrogen to oxygen is the same as their ratio in water, which is why glucose, starch, and cellulose are called carbo-hydrates. Small groups: Give students time to examine Figures 3.1, 3.2, and 3.3 carefully. To help students detect differences between cellulose and starch while recognizing that glucose is the monomeric unit of both, the models deviate somewhat from accepted conventions for representing carbohydrate structures. Note that the glucose monomers in starch are all oriented in the same way, whereas glucose monomers in cellulose alternate in their orientation. Science Note to Teacher: The piece of a starch molecule shown is amylopectin. Starch typically consists of both amylopectin, which is branched, and amylose, which is a simple chain. Starch is primarily used for storage as a source of glucose for later use rather than for making up structures.

Some polymers, like starch, can also have branches coming off the main chain.

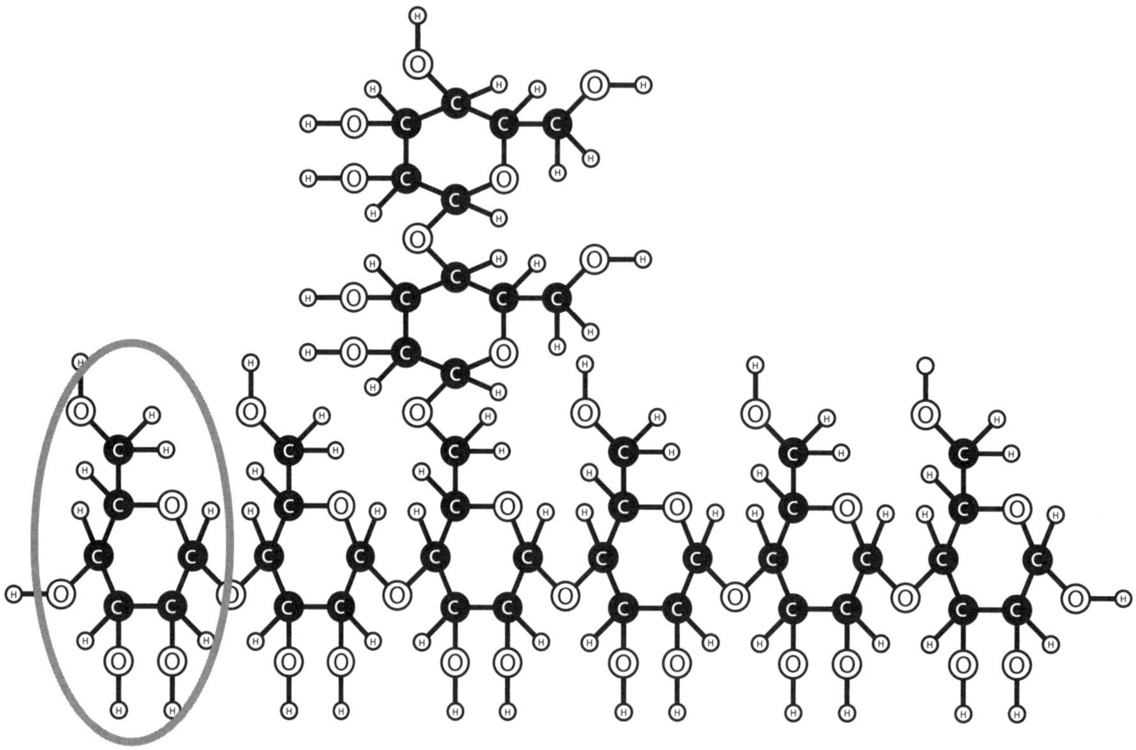

Figure 3.2. Model of the Structure of a Small Piece of a Carbohydrate Molecule Called ***Starch***

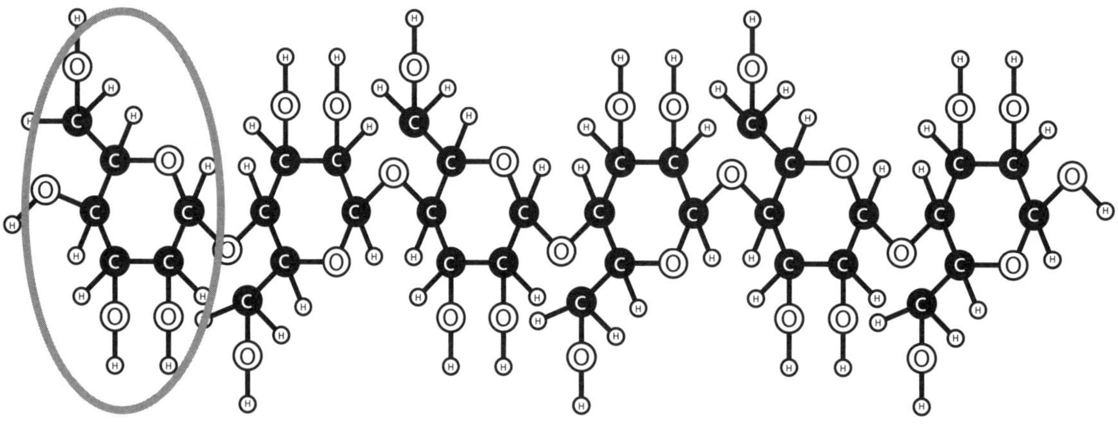

Figure 3.3. Model of the Structure of a Small Piece of a Carbohydrate Molecule Called ***Cellulose***

a. Look at the glucose molecule in Figure 3.1. Count the number of carbon, hydrogen, and oxygen atoms it is made up of, and use your counts to write the chemical formula for glucose. Hint: The atoms are in the order CHO in glucose.

$C_6H_{12}O_6$

Teacher Talk and Actions
Activity 2 (cont)
Lead a brief class discussion by asking some students to share their answers to the questions in Step 2.

b. Now look at the small piece of the starch molecule in Figure 3.2 and identify the repeating unit. Circle one repeating unit and write down how many units are shown in the piece of starch molecule. What did we call the repeating units of nylon? What did we learn to call a molecule made up of repeating units?

There are 8 glucose molecules in the piece of starch. A repeating unit is called a monomer. A polymer is made of monomers.

c. Now look at the piece of the cellulose molecule in Figure 3.3. Is cellulose also a polymer? If so, circle one repeating unit and write down how many are shown. Are they the same or different in their orientation (position)?

There are 6 glucose molecules in the piece of cellulose. Cellulose is a polymer. The first monomer is right side up, and the next one is flipped upside down—then the glucose monomers keep alternating up and down.

d. Do you think the glucose molecule in Figure 3.1 could be the monomer used to build the carbohydrate polymers starch and cellulose? Why or why not?

Yes, because although a few atoms would rearrange, it has nearly the same structure as the repeating units in starch and cellulose.

Teacher Talk and Actions
Activity 2 (cont)
Step 2d gives students practice in orienting to the models. Being able to recognize the patterns in the 2-D models will prepare them for using 3-D models later in the chapter.

Science Ideas

Lessons 3.1 and 3.2 were intended to help you understand some important ideas about carbohydrate molecules. Read the idea below. Remember that because science ideas are consistent with evidence from a wide range of phenomena, you are justified in using them to explain similar phenomena. Look back through Lessons 3.1 and 3.2 and describe evidence that illustrates each part of the science idea.

Science Idea #11: Carbohydrates are the main polymers making up plant body structures. Carbohydrate polymers are molecules made of glucose monomers. Different carbohydrate polymers have different properties because they are made of different numbers and arrangements of glucose monomers.

Evidence:

Table 2.1 in Lesson 3.1 provides data that show that plant parts, such as carrot roots, celery stems, and spinach leaves contain more carbohydrates than proteins or fats.

In Activity 1 of Lesson 3.2, we looked at three different carbohydrate molecules and saw that they have different properties from each other. For example, glucose is soluble in water while starch and cellulose are not, and cellulose forms fibers while starch and glucose do not. This means all three carbohydrates must be different substances.

In Activity 2 of Lesson 3.2, we saw the structures of two carbohydrate polymers, starch and cellulose, that are made from glucose monomers. Though a few atoms are arranged differently at the connections, it is easy to see the ring structure of glucose repeated in different patterns in these two polymers.

Teacher Talk and Actions
Science Ideas **(3 min)**
Read the science idea aloud to the class. Give students time to write down at least one example that supports the science idea. Remind students that the science ideas will be used in constructing scientific explanations. Add the science idea to the running classroom list of ideas and ask students to identify other related science ideas on the list. Students should notice that Science Idea #11 is an example of Science Idea #1.

Pulling It Together

Answer individually and be prepared for a class discussion.

1. Now that you know more about carbohydrates, how would you answer the Key Question, **Are the carbohydrates that make up different plant structures the same substance?** Provide as much evidence as you can.

Different plant structures probably have different amounts of cellulose, starch, and glucose. Structures like stems that need to be tough could have more cellulose, which is made up of fibers. Structures like roots that don't need to be so strong, could have a higher percentage of starch.

2. What chemical reaction do you think might happen when plants build cellulose polymers? Write the word equation for your proposed reaction below, and give a reason for why you think so.

glucose monomers → cellulose polymer + water molecules

I think this might be correct because it is like other examples of polymer formation we have seen. When nylon is formed from (hexamethylenediamine and adipic acid) monomers water molecules are produced.

SE p. 115

Teacher Talk and Actions

Pulling It Together **(7 min)**

Individuals: Students answer Pulling It Together questions on their own, in detail.

Whole class: Lead a class discussion.

Invite students to share ideas about Pulling It Together Question 1, which returns to the lesson Key Question. Direct students to look back at their initial ideas about the Key Question and consider how their ideas have changed.

Students should have an improved answer for Pulling It Together Question 1, the Key Question.

Remind students that they do not need chemical formulas in word equations.

Closure and Link **(3 min)**

Whole class:

Summarize the lesson in a few statements:

We saw that carbohydrate polymers are made up of glucose monomers. We saw that different carbohydrates have different properties and are therefore different substances.

Link to the next lesson:

In the next lesson, we will investigate how plants get the glucose monomers they need to build their body structures, which are mostly carbohydrate polymers like cellulose.

Lesson Guide
Chapter 3, Lesson 3.3
Making Glucose in Plants

Focus: This lesson focuses on the chemical reaction that plants use to make glucose (photosynthesis). Students reason about the substances that plants take in that could react to form glucose, model the reaction, and see experimental evidence about which reactant contributes the most to the mass of glucose.

Key Question: How can plants get the glucose monomers they need for building carbohydrate polymers?

Target Idea(s) Addressed

- Plants use carbon dioxide and water in their environment to make glucose monomers and oxygen molecules.

Science Practice(s) Addressed

- Analyze and interpret data to provide evidence for phenomena.
- Use multiple models to describe phenomena.
- Apply scientific ideas, principles, or evidence to construct an explanation for real-world phenomena.

Materials

Activity 3: Ball-and-stick model kits (1 per group of 3–4 students), Copymasters: *Chemical Reaction Mat:* Making Glucose, *Model Key,* Trays for working with models on desks

Advance Preparation

Practice modeling the reaction on the *Chemical Reaction Mat:* Making Glucose so that you will be able to quickly help students correct their models when needed.

Organize materials in advance so that each set of partners or table teams has materials when class begins.

Phenomena, Data, or Models	Intended Observations	Purpose	Rationale or Notes
Activity 1 Data on substances plants take in from their environment (water, carbon dioxide, and minerals).	Plants take in water, carbon dioxide, and minerals. Most of the mass of a tree is carbon, hydrogen, and oxygen atoms. Carbon dioxide and water are sources of carbon, hydrogen, and oxygen atoms. The amount of minerals plants take in does not contribute much mass.	Students use data to support their reasoning that carbon dioxide and water are likely reactants to form glucose. They should also understand that, while minerals are needed, they are only taken in small amounts and therefore aren't major contributors to the mass increase in plant growth.	The data are intended to address a common misconception: that most of the increase in a plant's mass comes from minerals in the soil. The data should help convince students that while minerals are taken in by plants, they make up a very small amount of plant bodies.
Activity 2 Data from isotopic labeling experiments that show where carbon from carbon dioxide ends up and where oxygen from water ends up as a result of the photosynthesis reaction.	Observe: Carbon from CO_2 ends up in glucose. Observe: Oxygen from water ends up in oxygen gas. Infer: Oxygen from CO_2 ends up in glucose. Conclude: The C and O atoms in glucose came from carbon dioxide.	Students are introduced to the basics of a commonly used biochemical method for "tracing" atoms through chemical reactions that happen in living organisms. (This method will appear in several activities in Chapters 3 and 4.) Students have the opportunity to analyze and reason from data to support the claim that the C and O atoms of glucose come from CO_2.	Isotopic labeling experiments, a method not typically encountered at the middle school level, is a powerful tool for understanding chemical reactions in living things. The lesson does not explain how scientists track atoms or how they know which molecules the atoms are a part of. The reasoning and inferences here are sophisticated, but the intention is for students to continue to let go of the misconceptions that gases don't contribute to growth and that nonliving matter and living matter are separate.

Phenomena, Data, or Models	Intended Observations	Purpose	Rationale or Notes
Activity 3 Ball-and-stick modeling activity that represents the (overall) reaction that plants use to convert carbon dioxide and water into glucose monomers and oxygen (photosynthesis)	A chemical reaction happens inside plants in which water reacts with carbon dioxide to form glucose monomers and oxygen molecules. This chemical reaction is called photosynthesis, and it explains where plants get the monomers they need for growth.	Students use models to visualize the chemical reaction that happens inside plants to produce glucose monomers for growth.	This is a simplification of the process of photosynthesis; students model the net reaction.
Pulling It Together A plant grown in a room enriched with added carbon dioxide will grow bigger than a plant grown in a room with normal air.	If students build models of the number of carbon dioxide molecules in the two rooms (7 in one and 18 in the other) and plenty of water molecules (18 in each room is fine) and then react the carbon dioxide and water molecules to form glucose, they should observe that 1 glucose molecule can be made in the first room and 3 glucose molecules can be made in the second room.	Students use models to support the claim that plants grown with added carbon dioxide will grow bigger. The models show that it is possible to build more glucose if more carbon dioxide (and sufficient water) is present. Reinforces the role that models can play in supporting explanations.	Students should have concluded from Lesson 3.2, Activity 2, that both starch and cellulose could have been formed from glucose monomers. Some students might also notice that the cotton polymer in Lesson 1.6 looks like it could have been made from glucose monomers.

Lesson 3.3—Making Glucose in Plants

What do we know and what are we trying to find out?

We now know that plant body structures are made up mostly of carbohydrate polymers like cellulose and starch. The molecular structures of these polymers suggest that they could be made from glucose monomers. But where do the glucose monomers come from? Do plants "eat" glucose monomers or do they make them? It turns out that there is almost no glucose in the soil or water. Bacteria in the soil and water devour any glucose that finds its way there. So if plants need glucose monomers, they probably have to make them. This lesson will help you figure out how.

Answer the Key Question to the best of your current knowledge. Be prepared to share your ideas with the class.

Key Question: How can plants get the glucose monomers they need for building carbohydrate polymers?

Maybe they make the glucose molecules from things they take in, like water and minerals/nutrients/soil. It probably involves a chemical reaction because glucose is a different substance from the substances I think that plants take in.

Teacher Talk and Actions
What do we know and what are we trying to find out? (3 min)
Whole class: Read the text under "What do we know and what are we trying to find out?" aloud to the class.
Key Question (5 min)
Individuals: Students respond to the Key Question: **How can plants get the glucose monomers they need for building carbohydrate polymers?** Whole class: Lead discussion. Invite student volunteers to share ideas about the Key Question. The goal is primarily eliciting and probing student ideas here. When students share their answers, do not indicate if their answers are right or wrong. However, do not let students assume by your silence that they are correct. Challenge any ideas that are inconsistent with learning from previous lessons.

Activity 1: Substances That Plants Take In From Their Environment

In the previous lesson, we saw that glucose is made up of carbon, hydrogen, and oxygen atoms. But where do these atoms come from? To figure that out, we might start by looking at what substances plants take in from their environment that could serve as a source of these atoms.

Procedures and Questions

1. Examine the information in Table 3.5 and answer the questions that follow.

Table 3.5. Substances Plants Take In From the Environment

Substance	Atomic Composition	Location in Environment	Plant Structure That Takes It in
Water (H_2O)	H, O	Ground	Roots
Carbon dioxide (CO_2)	C, O	Air	Tiny holes in the leaves
Minerals (in various molecular forms)	Mainly N, K, Ca, P, Mg, S*	Ground	Roots

**N = Sodium, K = Potassium, Ca = Calcium, P = Phosphorus, Mg = Magnesium, S = Sulfur*

a. Based on what you know about the molecular composition of glucose, which of the substances in the table could supply the needed atoms?

Water and carbon dioxide could, because they are made up of C, O, and H atoms, and so is glucose. A chemical reaction could happen where the atoms from CO_2 and H_2O rearrange to form molecules of glucose.

b. Will any of the minerals plants take in from the environment become part of glucose? Why or why not?

No, because minerals are not made of C, H, and O atoms.

SE p. 117

Teacher Talk and Actions	
Activity 1	**(10 min)**

Small groups: Give students time to read Table 3.5 and and work through Step 1.

This activity is intended to challenge the common misconception that most of the mass of a plant comes from minerals from the soil.

Watch student writing and oral explanations carefully for these misconceptions.

Science Note to Teacher:

Table 3.5 simplifies the way plants take minerals in. Plants usually get their minerals in various molecular forms, such as KNO_3 and NH_3 (not as single atoms), but showing these molecular structures would have complicated the student story line.

Activity 2: Can Plants Make Glucose From Carbon Dioxide and Water?

In the last activity, you decided that carbon dioxide and water molecules could supply plants with the atoms needed to make glucose. In this activity, you will use data from two experiments that scientists carried out to determine whether plants actually do make glucose from carbon dioxide and water. The experiments were first carried out in the 1940s using green algae. Later experiments gave the same results when carried out with plants that live on land or in the water.

Procedures and Questions

1. The experiments used a technique called *isotopic labeling* to track atoms from starting substances to ending substances.

   Isotopic labeling takes advantage of the fact that some atoms of the same type happen to be a little different from the regular atoms. These different "versions" of the atoms (called *isotopes*) have different masses but behave the same in chemical reactions. Some isotopes are unstable (radioactive) and break apart into fragments that can be detected with special instruments. For example, an unstable isotope of carbon has a mass of 14 amu and is referred to as carbon-14 or ^{14}C. Other isotopes are stable and must be detected by their masses. For example, a stable isotope of an oxygen atom has a mass of 18 amu. What do you think it is called?

   You will learn more about isotopes of atoms in high school chemistry, but two things are important for now. First, isotopes of atoms undergo the same chemical reactions as the regular versions, so they rearrange to form the same new molecules just like the regular versions do. Second, scientists can prepare molecules of starting substances that have some of the isotopic atoms. This allows scientists to use their instruments to see which molecules have the isotopes at the end of the chemical reaction. In this way, scientists can figure out what the starting and ending substances are for chemical reactions that are ocurring in the bodies of living organisms.

   Let's look at two experiments that scientists carried out to try to find out where the carbon, hydrogen, and oxygen atoms of glucose come from. For each experiment, look carefully at the data in the table, and then work with your group to answer the questions.

Teacher Talk and Actions

Activity 2 **(40 min)**

Whole class: Read the first paragraph under Step 1 aloud to the class. Ask students what they think an oxygen atom with a mass of 18 amu would be called (Oxygen-18 or ^{18}O). [If a student asks what the mass of "normal" oxygen is, you can remind them of Table 2.3 in Lesson 2.2, and say that normal oxygen has a mass of 16 amu. However, don't digress if no one asks.]

Continue reading the next paragraph. Pause frequently to make sure students are following—isotopic labeling is a crucial idea throughout the rest of Chapters 3 and 4.

Read the last paragraph to introduce the experiments.

Students should observe from the passage:

- Some atoms have isotopic "versions."
- Scientists can incorporate isotopic versions of atoms into molecules they want to study.
- Isotopically labeled atoms allow scientists to track atoms of reactants to see where they end up during chemical reactions.

Science Note to Teacher:

^{14}C is an unstable isotope, whose presence is detected by its decay products. Because it decays, it is called a *radioactive isotope.* ^{18}O is a stable isotope that does not decay, so its presence is detected by its mass.

2. Experiment 1: In the first experiment, scientists used carbon dioxide made with labeled carbon atoms so that they would be able to see where the carbon atoms ended up after the carbon dioxide and water molecules reacted. Table 3.6 summarizes their data.

Table 3.6. Location of Labeled Atoms (highlighted) at the Beginning and End of Experiment 1

	Starting Substances (Reactants)		**Ending Substances (Products)**	
Experiment 1	CO_2	H_2O	$C_6H_{12}O_6$	O_2

a. What can scientists conclude from Experiment 1 about the carbon atoms? What piece of data can they cite as evidence for their conclusion?

Scientists can conclude that the C atoms in glucose came from carbon dioxide. The evidence they can cite is the observation that the labeled C atoms that started out in carbon dioxide ended up in glucose.

b. Does the data provide evidence for a conclusion about where the oxygen atoms from carbon dioxide end up? Why or why not?

No, because they didn't track the O atoms. The O atoms could have come from carbon dioxide or water. We don't know yet.

3. Experiment 2: In the second experiment, scientists used water made with labeled oxygen atoms and looked to see where the oxygen atoms ended up after water and carbon dioxide reacted. Table 3.7 summarizes their data.

Table 3.7. Location of Labeled Atoms (highlighted) at the Beginning and End of Experiment 2

	Starting Substances (Reactants)		**Ending Substances (Products)**	
Experiment 2	CO_2	H_2O	$C_6H_{12}O_6$	O_2

a. What can scientists conclude from Experiment 2? What piece of data can they cite as evidence for their conclusion?

Scientists can conclude that the O atoms in water are the same O atoms as are found in oxygen gas. They can cite Experiment 2 as evidence, and how the isotopically labeled oxygen was later found in oxygen gas.

Teacher Talk and Actions
Activity 2 (cont)
Small groups: Give students time to study Table 3.6 and answer Questions 2a and 2b.
Discuss student responses.
Question 2a. Some students might conclude that the C atoms of $C_6H_{12}O_6$ come from CO_2 because CO_2 has C atoms. However, that is not considered sufficient evidence. Instead, students should recognize that the observation that the C atoms of glucose have the label is stronger evidence.
Question 2b. Make sure students appreciate that without being able to track labeled atoms, in this case oxygen atoms, scientists won't know whether the O atoms come from CO_2, H_2O, or both.

b. What can scientists conclude about where the O atoms that make up glucose come from?

Because the only other input that has O atoms is carbon dioxide, the O atoms in glucose must come from carbon dioxide.

c. What evidence can they cite for this conclusion?

They can cite Experiment 2, because scientists know that plants only take in water and carbon dioxide. Because the O atoms didn't come from water, they have to come from carbon dioxide.

4. Write the word equation for the chemical reaction that plants use to make glucose.

carbon dioxide + water → glucose + oxygen gas

5. Write the chemical formulas for each starting and ending substance:

Starting Substances	Ending Substances
H_2O and CO_2	$C_6H_{12}O_6$ and O_2

6. Draw a diagram to illustrate the reaction in Experiments 1 and 2. Based on data from the experiments, use arrows to show where the C and O atoms are located in the reactants and where they end up in the products of the reaction.

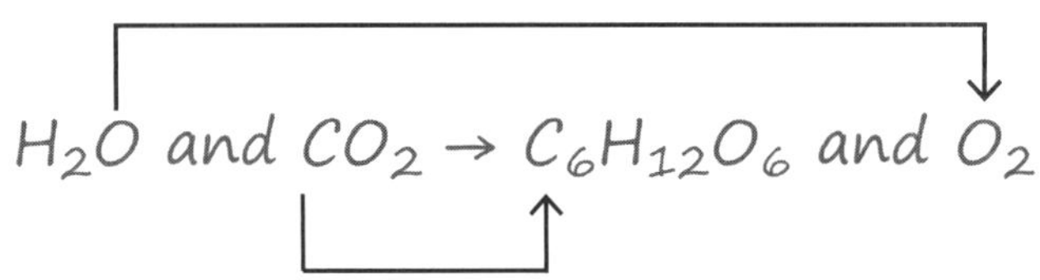

Teacher Talk and Actions

Activity 2 (cont)

Question 3b requires sophisticated reasoning.

If students struggle, let them know that neither experiment directly shows where the oxygen atoms in glucose came from. Tell them to focus on the two possible sources of oxygen atoms and use the process of elimination.

Step 6: The line should go from the O of H_2O to the O of O_2. This is what Experiment 2 shows. Students can infer, as scientists did, that if H_2O provides the O atoms for O_2, then the O atoms making up $C_6H_{12}O_6$ probably come from CO_2.

If students struggle with this step, in Activity 3 they can use different colored stickers to help them figure it out. They can use one color to label some of the carbon atoms and a different color to label some of the oxygen atoms on the water molecule they build.

Students shouldn't take this experiment as providing the best possible evidence. Teachers can ask students what experiment could be done to provide better evidence that the O atoms from glucose come from CO_2. Students should appreciate that they would grow the plants with $C^{18}O_2$.

Science Note to Teacher:

There is a seeming inconsistency that the net equation has six O atoms (from six H_2O) but needs to make 12 O atoms (in six O_2 molecules). The net reaction is the sum of two reactions: The light reaction requires an additional six H_2O (12 H_2O per six O_2), and the dark reactions produce six H_2O per glucose. See Chapter 3 Teacher Background Information for more information.

Activity 3: Modeling the Chemical Reaction That Plants Use to Make Glucose

Materials
For each team of students
Ball-and-stick model kit
Chemical Reaction Mat: Making Glucose
Tray for working with models on desks
Model Key

In the last activity, you determined that carbon dioxide and water are the starting substances for the chemical reaction that produces glucose and oxygen, and that carbon dioxide provides the C atoms for making glucose and probably the O atoms too. In this activity, you will figure out how many carbon dioxide molecules and how many water molecules are needed to make each glucose molecule. Then you will model the chemical reaction to see if glucose and oxygen molecules can be made by rearranging atoms from those carbon dioxide and water molecules. You should also be able to figure out how many oxygen molecules will be produced.

Procedures and Questions

1. Use the information in Table 3.6 of Activity 2 to figure out how many carbon dioxide molecules and water molecules are needed to make a molecule of glucose. Record your answers in the table below.

How many carbon dioxide molecules are needed to make a molecule of glucose?	6
How many water molecules are needed to make a molecule of glucose?	6

2. Examine the green side of your *Chemical Reaction Mat* to check your answers. Use your ball-and-stick model kit to build the reactant molecules. Depending on the size of your team, each member will need to make 3 or 4 molecules.

3. Close the ball-and-stick kit and put it aside.

4. Flip your mat to the yellow side.

5. Work together to build the product molecules on the yellow side of the mat from the reactant molecules.

6. Record the number of molecules your team has made:

Number of glucose molecules	1
Number of oxygen molecules	6

Teacher Talk and Actions

Activity 3 **(25 min)**

Pass out a model kit and the *Chemical Reaction Mat:* Making Glucose to each group of three to four students.

Remind students that this modeling activity will help them answer the Key Question: **How can plants get the glucose monomers they need for building carbohydrate polymers?**

Small groups: Give students time to work through Steps 1–3. Encourage students to assign responsibilities for making molecules equitably among team members.

Step 1: Students should reason from the formula for glucose that 6 C are needed since each carbon dioxide molecule has 1 C atom. Students should reason that 12 H atoms are needed, and since each water molecule provides 2 H atoms, 6 are needed as well.

For logistical reasons, student must build their own carbon dioxide and water molecules. Reinforce that this stage is just preparing the starting substances that they will then react to form the ending substances.

Once they've made their reactants, circulate and check that they have built their molecules correctly.

Step 5: Note that the models of reactant CO_2 molecules have 24 curved sticks, representing the double bonds linking C to O. However, the models of the product O_2 molecules only have 12 curved sticks, representing the double bonds between the O atoms. That means that students will need to use 12 curved sticks when building glucose. Some students find this problematic, because it doesn't quite look like the model of glucose on the *Chemical Reaction Mat.* Wait to see if students find this troublesome. If they do, here are two options:

- Tell students to use straight sticks to link C atoms and use the curved sticks to link C to O or O to H.
- Tell students that they can swap 12 curved sticks for 12 straight sticks in their kits.

Science Note to Teacher:

In this activity, students model a simplified version of photosynthesis—the net equation. See the Chapter 3 Teacher Background Knowledge for more information.

7. With your team, answer the following questions.

   a. Write the word equation for the reaction you just modeled. This chemical reaction is called *photosynthesis*.

      carbon dioxide + water → glucose + oxygen gas

   b. Write the equation for the chemical reaction you modeled, using the chemical formula for each reactant and product. In front of each reactant, write how many molecules of it were involved in the reaction. Then in front of each product, write how many were produced.

      $6\ CO_2 + 6\ H_2O \rightarrow 1\ C_6H_{12}O_6 + 6\ O_2$

   c. Are atoms rearranged during this chemical reaction? Use models to support your answer.

Yes. According to the model of a glucose molecule in Figure 3.1 on p. 111, glucose contains carbon atoms linked to other carbon atoms. In carbon dioxide, a carbon atom is linked to two oxygen atoms, not to other carbon atoms. So carbon atoms are in different arrangements. The same is true for oxygen atoms. Oxygen gas has two O atoms linked together, whereas in the water molecule O is linked to two H atoms. So O atoms disconnect from H atoms and connect to one another.

   d. How does the number of carbon, hydrogen, and oxygen atoms in the reactants compare with the number of carbon, hydrogen, and oxygen atoms in the products? Are atoms conserved during this chemical reaction?

There are the same number of carbon, hydrogen, and oxygen atoms in the products as in the reactants. For example, there are 6 carbon in the products and 6 in the reactants. Atoms are conserved.

8. Carefully take apart your models and put the pieces back in the bag for the next class to use.

Teacher Talk and Actions
Activity 3 (cont)
The term *photosynthesis* is introduced after students have modeled the reaction—some students may already know it, and it will be a useful shorthand way to refer to "the reaction that plants use to make glucose." Question 7c prompts students to notice that before the reaction, no carbon atoms were linked together, but after the reaction, there are six carbon atoms linked in a chain/ring in glucose.

Science Ideas

Lesson 3.3 was intended to help you understand an important idea about how plants make glucose. Read the idea below. Remember that because science ideas are consistent with a wide range of relevant evidence, you are justified in applying them to new observations and data. Look back through Lesson 3.3. Give at least one piece of evidence from your work so far and explain how it supports this idea.

Science Idea #12: Plants use carbon dioxide and water molecules in their environment to make glucose and oxygen molecules. Atoms are rearranged and conserved during this chemical reaction.

Evidence:

In Activity 1, we saw that plants take in carbon dioxide and water from their environment, and Activity 2 showed that these could react to form glucose and oxygen. In Activity 3, we modeled the reaction and saw that when this reaction happens, the number of C, H, and O atoms remains the same even though they are arranged differently.

Teacher Talk and Actions
Science Ideas **(3 min)**
Read the science idea aloud to the class. Give students time to write down at least one piece of evidence that supports the idea. Remind students that they can now use this science idea when constructing scientific explanations. Add the science idea to the running classroom list of ideas.

Pulling It Together

1. Now that you know more about glucose, how would you answer the Key Question, **How can plants get the glucose monomers they need for building carbohydrate polymers?**

Plants take in carbon dioxide and water from their environment, which react inside plants to form glucose monomers and oxygen molecules.

2. Recall the reaction that formed rust. How is photosynthesis similar to the formation of rust? Think about a plant's body as being like the container that held the steel wool. Write about what you think happens to both measured mass and total mass of the plant.

Photosynthesis is similar to the formation of rust:

- *Both are chemical reactions in which new substances form because atoms of molecules of reactants rearrange to form molecules of products.*
- *Both chemical reactions have a gas as a reactant that can enter an open container.*
- *In both cases the measured mass increased in open systems. This is because in both cases atoms (making up a gaseous reactant) entered the container and reacted to form a product that "trapped" the atoms in molecules that were not gases.*
- *In both cases the total mass stayed the same (was conserved) even though the measured mass increased, because the increase in atoms inside the container was balanced by the decrease in atoms outside the container.*

Teacher Talk and Actions
Pulling It Together **(45 min)**
Individuals: Students answer Pulling It Together questions on their own, in detail. Whole class: Lead a class discussion. Invite students to share ideas about Pulling It Together Question 1, which returns to the lesson Key Question. Direct students to look back at their initial ideas about the Key Question and consider how their ideas have changed. Students should have an improved answer for Pulling It Together Question 1, the Key Question. Pulling It Together Question 2 draws the analogy between iron rusting and photosynthesis. Both involve an increase in mass when the reaction is carried out in an open system, because atoms from a gas rearrange to become part of a solid. An important difference between these two reactions is the role of oxygen: Oxygen is a reactant in the iron rusting reaction and it is a product of photosynthesis. Acknowledge this difference if students notice it. Students will learn how these two chemical reactions are related in high school chemistry.

3. The owner of a plant store wanted to see if the amount of carbon dioxide in the air would affect how her plants grow. She set up two identical rooms, Room 1 and Room 2, in her greenhouse. She put one plant in each room. Both plants were the same size and mass, and both were given plenty of water. Room 1 had normal air with the normal level of carbon dioxide in it. Room 2 was filled with air that had extra carbon dioxide in it. See Figure 3.4 below, or the color version your teacher projects.

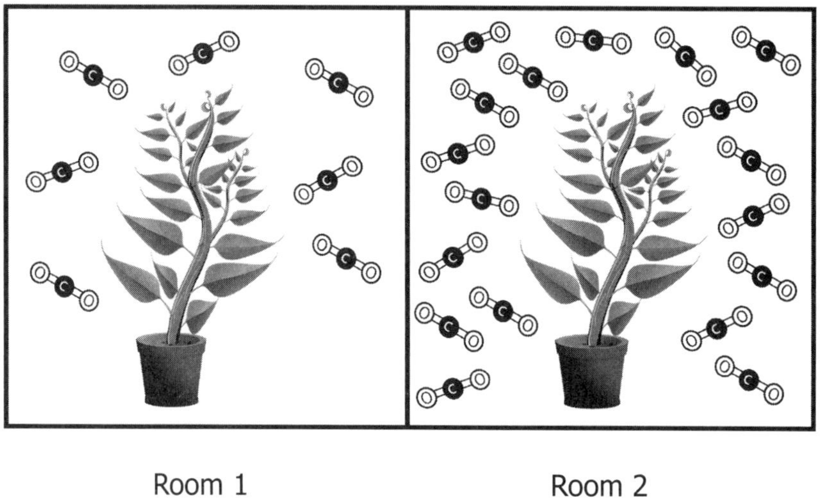

Figure 3.4. Plants Exposed to Different Amounts of Carbon Dioxide

Three months later, the store owner measured each of the plants. Do you think one plant was bigger than the other? Explain your answer using science ideas, evidence, and models. Check to be sure that your explanation meets the Explanation Quality Criteria.

The plant in Room 2 would probably be bigger. Plants use carbon dioxide and water molecules in their environment to make glucose and oxygen molecules (Science Idea #12). We are told that both plants have plenty of water, so we know that the amount of water won't limit the amount of glucose that the plants can produce. However, the plant in Room 2 has more carbon dioxide than the plant in Room 1, so the plant in Room 2 can make more glucose. When we modeled the chemical reaction with ball-and-stick models, we saw that we will use 6 molecules of CO_2 and 6 molecules of H_2O for each $C_6H_{12}O_6$ molecule we make. The drawings above (another type of model) show that there are 7 CO_2 molecules in Room 1 and 18 CO_2 molecules in Room 2. So if each room has at least 18 H_2O molecules then here's what will happen according to our models:

- *The plant in Room 1 will use 6 CO_2 and 6H_2O to make 1 $C_6H_{12}O_6$ and 6 O_2. There will be 1 CO_2 and 12 H_2O left over, because there isn't enough CO_2 to make another glucose molecule.*
- *The plant in Room 2 will use 18 CO_2 and 18 H_2O to make 3 $C_6H_{12}O_6$ and 18 O_2.*

The models show how it is possible for the plant in Room 2 to produce more glucose, which it can probably use to make more polymers to build its body structures.

Teacher Talk and Actions

Pulling It Together (cont)

Pulling It Together Question 3 is a challenging question for students to think through. A few students may be able to reason abstractly about why the plant in Room 2 will get bigger. Most will need help.

Have students work in small groups and encourage them to use LEGO or ball and stick models to help then think through their explanations. Multiple LEGO kits will have to be shared among small groups (one kit does not have enough bricks for this question). You can also let the more advanced students use ball and stick models.

You can give them the following hints as needed:

• Tell students that the drawings in Figure 3.4 are also models that they can use as a guide. (Remind them that real CO_2 molecules are much smaller and there are lots more in the rooms. However, the drawings give them a sense of the relative amounts of CO_2 in the two rooms).

• Start by building models of CO_2 molecules in each room, using the drawing as a guide.

• Then have them build models of H_2O molecules in each room. Check to be sure students have at least 18 H_2O molecules in each room.

• Then have students build glucose molecules that can be formed in each room from the CO_2 and H_2O molecules they have built. When building glucose, all that matters is the atomic composition.)

Once students have figured out their explanations with their models, have them construct a written explanation.

Have blank explanation charts available for those who still need them, which can be found on page xviii of the introduction to the unit.

Closure and Link (3 min)

Whole class:

Summarize the lesson in a few statements:

We saw that in plants, carbon dioxide gas from the air and water from the ground react to form glucose monomers, which remain in the plant, and oxygen molecules, which are released into the air.

Link to next lesson:

In the next lesson, we will investigate how plants make the carbohydrate polymers that make up their body structures from these glucose monomers.

Lesson Guide
Chapter 3, Lesson 3.4
Making Cellulose Polymers in Plants

Focus: This lesson focuses on the chemical reaction that plants use to make cellulose from glucose monomers. Students see experimental data from radioactive labeling experiments that show that plants can make cellulose from glucose, and then they model this reaction.

Key Question: How do plants make the cellulose polymers they need to build their body structures?

Target Idea(s) Addressed

- Plants use glucose monomers to make carbohydrate polymers and water molecules. Atoms are rearranged during this chemical reaction.
- When plants grow or repair, they increase in mass. Atoms are conserved when plants grow: The increase in measured mass comes from the incorporation of atoms from molecules that were originally outside of the plants' bodies.

Science Practice(s) Addressed

- Analyze and interpret data to provide evidence for phenomena.
- Use multiple models to describe phenomena.

Materials

- Activity 1: Projector, Slide show: *Herbicides*, Ball-and-stick model kits (1 per group of 3–4 students), Copymaster: *Chemical Reaction Mat:* Making Cellulose, *Model Key*, Trays for working with models on desks

Advance Preparation

Practice modeling the reaction on the *Chemical Reaction Mat:* Making Cellulose so that you will be able to quickly help students correct their models when needed.

Practice breaking the cellulose model down into glucose molecules. You will have to do this in between each class.

Build two glucose molecules for each team of 3–4 students to be distributed at the beginning of Activity 2. You will only need enough glucose molecules for your first class of the day. For the remaining classes you will break down the cellulose models that students build, which will give you the glucose monomers to distribute to the next class.

Organize materials in advance so that each set of partners or table teams has materials when class begins.

Phenomena, Data, or Models	Intended Observations	Purpose	Rationale or Notes
Activity 1 Data from a radioactive labeling experiment that show how many labeled atoms from glucose end up in cellulose in plants with and without herbicide application	Carbon atoms from glucose end up in cellulose. A reaction must be happening in which glucose monomers react to make cellulose. Less glucose reacts to form cellulose in plants treated with herbicide.	Students use experimental data to see evidence (by tracking atoms) that plants make cellulose from glucose.	The labeling method provides a way to give students evidence that what they have predicted (that cellulose is made from glucose) actually happens inside plants as they grow.
Activity 2 Ball-and-stick modeling activity to show how plants could build cellulose from glucose	Models show that cellulose polymers and water molecules can be made by reacting glucose monomers.	Students use models to visualize the reaction that plants use to build the polymer cellulose from glucose monomers.	This is the final piece in the plant growth puzzle.
Pulling It Together Although nonliving matter like rocks, water, and air don't look like living things like people, trees, and dogs, the C, H, and O atoms that make up nonliving things can be used to build body structures of plants.	Data provided in Chapter 3 provides the following evidence: • Plants take in non-living matter (carbon dioxide and water). • Plants can make glucose from atoms making up carbon dioxide and water molecules. • Plants can make cellulose from glucose monomers. • Plant body structures are mostly carbohydrates, like cellulose.	Students identify evidence from earlier activities to support their claim.	This directly confronts the misconception that living things are made of different matter than nonliving things.

Lesson 3.4—Making Cellulose Polymers in Plants

What do we know and what are we trying to find out?

We know that plant body structures are mostly made up of carbohydrate molecules, and that some carbohydrate molecules, like cellulose, are better for building body structures than others. When we examined the model of a piece of cellulose, we thought it might be made up of glucose monomers. But while models can give us ideas about how something might work, they don't provide evidence for claims. If we want evidence that plants actually make cellulose polymers from glucose monomers, we need to collect data. In this lesson, we will examine experimental data to see if it supports our claim or idea that plants make cellulose polymers from glucose monomers.

Answer the Key Question to the best of your current knowledge. Be prepared to share your ideas with the class.

Key Question: How do plants make the cellulose polymers they need to build their body structures?

Maybe plants can use glucose monomers to make cellulose polymers (and water), sort of like making nylon polymers (and water molecules) from hexamethylenediamine and adipic acid monomers.

Teacher Talk and Actions
What do we know and what are we trying to find out? (3 min)
Whole class: Read the text under "What do we know and what are we trying to find out?" aloud to the class.
Key Question (5 min)
Individuals: Students respond to the Key Question: **How do plants make the cellulose polymers they need to build their body structures?** Whole class: Lead discussion. Invite student volunteers to share ideas about the Key Question. The goal is primarily eliciting and probing student ideas here. When students share their answers, do not indicate if their answers are right or wrong. However, do not let students assume by your silence that they were correct. Challenge any ideas that are inconsistent with learning from previous lessons.

Activity 1: Following "Labeled" Carbon in Glucose

In this activity, you will again take advantage of the isotopic labeling method to help you figure out whether growing plants use glucose to make cellulose polymers. In the experiment described below, scientists were trying to stop plants from growing. However, the data they collected also provide evidence for how growing plants use glucose to build their body structures.

Procedures and Questions

1. Watch as your teacher projects information about herbicides.
2. Read about the experiment that scientists conducted, examine the data they collected, and then answer the questions that follow with your group.

Effects of the Herbicide Triazofenamide on Plant Growth

Introduction

Herbicides are substances that stop or prevent the growth of weeds (unwanted plants). In agriculture, weeds take away resources like water and space from desired plants or crops. Herbicides prevent weeds from growing, so farmers often spray their fields with them. There are many different herbicides that are commonly used. Scientists wanted to know if the herbicide triazofenamide also stopped the growth of the surrounding plants.

Experiment

Mouse-ear cress plants were split into two groups: **experimental** and **control**. The experimental group was sprayed with the herbicide, and the control group was not. Although plants normally make their own glucose, scientists gave the plants glucose molecules that had labeled carbon atoms, and then measured how many of the labeled carbon atoms ended up in cellulose in both experimental and control plants. The amount of labeled carbon atoms in cellulose would tell scientists how much cellulose was produced by each group.

Results

Table 1. Amount of labeled carbon atoms in cellulose in dpm (a unit that indicates the level of radioactivity in a substance or material) for each group in the experiment

Group	Amount of Labeled C Atoms in Cellulose (dpm)
Control	11,200
Experimental (herbicide)	1,900

Heim, D. R., I. M. Larrinua, M. G. Murdoch, and J. L. Roberts. 1998. Triazofenamide is a cellulose biosynthesis inhibitor. *Pesticide Biochemistry and Physiology* 59 (3): 163–168.

Teacher Talk and Actions	
Activity 1	**(15 min)**

Read the text under Activity 1 aloud to your class.

Project the slide show *Herbicides* and read the Background, Experimental Question, and Experiment sections aloud to the class.

Small groups: Give students time to analyze the data in Table 1 and answer Questions 2a–d.

Science Note to Teacher:

dpm stands for "disintegrations per minute" and is a measure of radioactivity. There is no need for students to focus on this unit of measurement—it was included so that the actual data could be shown, which were more powerful than just abstractions like "more" or "x many times more."

Some students have the misconception that the major source of a plant's food is the glucose that it takes in through the roots from the soil. However, while plants actually get the bulk of their glucose by making it via photosynthesis, plants can take up glucose if it is available. However, the soil in nature rarely has much glucose in it because microorganisms consume it all.

a. Based on the data in Table 1, can mouse-ear cress plants make cellulose from glucose? What evidence supports your answer/claim?

Yes, plants can make cellulose from glucose because labeled atoms from glucose ended up in cellulose in both the control and the experimental groups (though more ended up in cellulose in the control group).

b. Do the data show that the herbicide prevents mouse-ear cress plants from making cellulose from glucose? Why or why not?

Yes, this experiment showed that the herbicide mostly prevented the plant from making cellulose from glucose because the plants that got the herbicide (the experimental group) made much less cellulose than the plants that didn't have the herbicide (the control group).

c. Do the data show that the herbicide prevents mouse-ear cress plants from carrying out photosynthesis (making glucose)? Why or why not?

No, they don't, because the scientists gave the plant glucose, so we could not know from this experiment if the herbicide would prevent plants from making glucose.

d. How do you predict the mass of the experimental plants would compare with the mass of the control plants after several weeks? Explain your prediction.

I predict the body structures of the experimental plants would not grow normally. Their structures might be smaller or not properly shaped. This is because plant body structures are made up of cellulose, a fibrous polymer that is not soluble in water. Glucose is soluble in water and is not fibrous, so glucose alone would not be enough for a plant to grow normal body structures.

Teacher Talk and Actions
Activity 1 (cont)
Lead a discussion by asking students to share their answers. Question 2a requires students to use an experiment that was designed to answer a different question to help answer their question: Can plants make cellulose from glucose? The control data are most useful for this, though both groups show at least some cellulose was made from glucose. Question 2b checks students' comprehension of what was manipulated and what was being tested in the experiment. Question 2c checks to see if students understand the answer that scientists got to their question: How does the herbicide stop plants from growing? Question 2d connects the molecules being made (or not made) to body structures and requires students to reason about the likely macroscopic effects of the herbicide on plants.

Activity 2: Modeling the Reaction That Plants Use to Build Cellulose

Materials
For each team of students
Two premade models of glucose molecules (with stickers on one or more carbon atoms)
Chemical Reaction Mat: Making Cellulose
Tray for working with models on desks
Model Key

So we now know that mouse-ear cress plants can produce cellulose polymers from glucose monomers. Do any atoms rearrange during the process? Let's model the reaction to get ideas about whether any atoms form different connections (rearrange) and, if so, which ones.

Procedures and Questions

1. With your group, look at the molecule models you have been given, and then look at the yellow side of the *Chemical Reaction Mat*. How many new connections will form?

2. On the green side of the mat, draw a line between the atoms that will connect to form the piece of cellulose polymer.

3. Flip your *Chemical Reaction Mat* to the yellow side.

4. Using only the reactant models, build models of the products as shown on the yellow side of the mat. **Do not take apart the models you have just built—leave them for the next class.**

5. With your team, answer the following questions.

    a. Did any atoms rearrange when you modeled this reaction? If so, which atoms rearranged?

    One hydrogen from one glucose monomer, and one hydrogen and one oxygen from another glucose monomer disconnected from the monomers. Then the H and OH connected to form water.

    b. Write a word equation for the reaction you just modeled.

    glucose monomers → cellulose polymer + water molecules

    c. When this reaction happens in a plant, what must be happening to the mass of the plant's body structures? Justify your answer.

    The mass must be increasing because new body structures are being formed from cellulose, and cellulose has mass.

Teacher Talk and Actions	
Activity 2	**(15 min)**

To each group of three to four students: Pass out two glucose molecules (that you made earlier) and *Chemical Reaction Mat:* Making Cellulose.

Small groups: Give students time to work through Steps 1–5.

Remind students not to take apart their models—circulate to be sure they are built correctly.

You will have to unhook the cellulose models between class periods so that the next class can start with two glucose molecules.

(Don't ask students to do this as it might be confusing for them to immediately undo the reaction they just did.)

For time and logistical reasons, students only model linking two glucose monomers. Remind them that cellulose is a long chain that would form by repeating this reaction many, many times.

Science Ideas

Lesson 3.4 was intended to help you understand some important ideas about how plants grow. Read the ideas below. Notice that Science Ideas #13 and #14 explain observations of plant growth in terms of atoms. We can observe when plants grow, but we can't see atoms rearranging or entering or leaving the system. However, because Science Ideas #13 and #14 state general principles that are consistent with a wide range of observations and data, we can use them to reason about the growth of all plants. You will be expected to use ideas about atoms to explain phenomena involving plant growth.

Look back through Lesson 3.4 and give at least one piece of evidence for each idea.

Science Idea #13: Plants use glucose monomers to make cellulose polymers and water molecules. Atoms are rearranged during this chemical reaction.

Evidence:

In Activity 1, we saw evidence that plants make cellulose from glucose when labeled carbon atoms from glucose turned up in cellulose. In Activity 2, we modeled this reaction and saw that atoms rearranged. We reasoned that the cellulose is being used to build more body structures, since most plant body structures are made up of cellulose.

Science Idea #14: When plants grow or repair, they increase in mass. This increase in measured mass comes from the incorporation of atoms from molecules that were originally outside of the plants' bodies.

Evidence:

In Lesson 3.4, we saw that plants make cellulose from glucose, and we saw earlier that they make glucose from carbon dioxide and water they take in from their environment. Plants increase in mass because atoms originally from carbon dioxide in the air and water in the ground are now incorporated into their body structures.

Teacher Talk and Actions
Science Ideas (5 min)
Read the science idea(s) aloud to the class. Give students time to write down at least one example that supports each idea. Remind students that the science ideas will be used in constructing scientific explanations. Add the science idea(s) to the running classroom list of ideas, noting similarities to other ideas. Science Idea #13 is an example of Science Ideas #2, #5, and #6. Science Idea #14 is an example of Science Ideas #7 and #10.

Pulling It Together

1. Now that you've learned more about cellulose, how would you answer the Key Question, **How do plants make the cellulose polymers they need to build their body structures?** Start with glucose.

In plant bodies, glucose monomers can react to form cellulose polymers (and water molecules). These cellulose polymers are used to build body structures during growth and repair.

2. A friend says that "living things, like people and trees and dogs, are very different from nonliving things like rocks, water, and air. Nonliving matter could never become part of a living thing." Is he correct or incorrect? What is your evidence?

The friend is incorrect. Chapter 3 provided the following evidence that plants are made up of some of the same atoms as nonliving things:

- *Data in Table 3.5 (p. 117) provided evidence that plants can take in H_2O and CO_2 molecules from their environment. Before plants take in these molecules, they are part of nonliving things.*
- *Data represented in Experiment 1 (Table 3.6, p. 119) provided evidence that plants made radioactively labeled glucose ($C_6H_{12}O_6$) from radioactively labeled CO_2. The carbon atoms that used to be part of nonliving things are now part of glucose.*
- *Data presented in herbicide experiment (p. 127) provided evidence that plants can make cellulose from glucose.*
- *Data presented in Table 3.2 (p. 105) provides evidence that plant body structures are mostly made up carbohydrates, which we later learned are mostly starch and cellulose.*

So the C atoms from CO_2 can become part of the cellulose that makes up plant body structures.

Teacher Talk and Actions

Pulling It Together (5 min)

Individuals: Students answer Pulling It Together questions on their own, in detail.

Whole class: Lead a class discussion.

Invite students to share ideas about Pulling It Together Question 1, which returns to the lesson Key Question. Direct students to look back at their initial ideas about the Key Question and consider how their ideas have changed.

Students should have an improved answer for Pulling It Together Question 1, the Key Question.

For Pulling It Together Question 2, there are at least four pieces of evidence. Give students time to find as many of the pieces of evidence as possible. During the discussion, as each piece of evidence is presented, ask, "What is this evidence of?"

3. How is plant growth similar to nylon formation?

Both:

- *Produce substances that have different properties from starting substances.*
- *Involve chemical reactions that form polymers (and water molecules) from monomers.*
- *Produce polymers that can be used to build fibers.*
- *Involve atom rearrangement: Atoms of molecules of starting substances (reactants) rearrange to form molecules of ending substances (products).*
- *Conserve matter because atoms are not created or destroyed.*

4. Look back at Lesson 3.2, Figure 3.2 (p. 112). How do you think plants might make starch molecules, starting with glucose molecules? *Maybe plants could make starch polymers like they make cellulose polymers: Glucose monomers could react to form long chains (and water molecules). Based on Figure 3.2, starch polymers have branches, so plants would need to carry out different chemical reactions to build branches off the chains. Also, it looks like the glucose monomers are linked differently in cellulose compared with starch polymers, so plants would need to be able to link the glucose monomers differently to form starch versus cellulose.*

Teacher Talk and Actions

Pulling It Together (cont)

Pulling It Together Question 3 requires students to identify similarities between reactions in nonliving and living things. Challenge students to identify as many as they can. More advanced students can also be asked how the reactions differ (different monomers).

Pulling It Together Question 4 requires students to apply what they have learned throughout the unit about polymers to reason (without seeing direct evidence) about how starch could form. Students should use words like "maybe" and "probably."

If students don't notice the difference in how glucose molecules are linked in starch versus cellulose, focus their attention on the third and fourth glucose monomers: In starch each glucose molecule has two OH groups pointing down whereas in cellulose the two OH groups point down in the third glucose and they point up in the fourth glucose.

Closure and Link (2 min)

Whole class:

Summarize the lesson in a few statements: We saw that in plants, glucose monomers (which were made from carbon dioxide and water) can react to form cellulose polymers (and water). Cellulose is what plant body structures are mostly made up of, so the addition of cellulose explains how plants grow and repair.

Link to next lesson:

In the next lesson, we will summarize our ideas about how atoms and molecules help us understand plant growth.

Lesson Guide
Chapter 3, Lesson 3.5
Explaining Where the Mass of Growing Plants Comes From

Focus: This lesson does not introduce new ideas but rather is intended to assess how well students understand plant growth and whether they can integrate ideas from Chapters 1 and 2 with what they have learned in Chapter 3.

Key Question: Where does the mass of growing plants come from?

Target Idea(s) Addressed

- Students should be able to use all the science ideas from Chapters 1, 2, and 3.

Science Practice(s) Addressed

- Analyze and interpret data to provide evidence for phenomena.
- Apply scientific ideas, principles, or evidence to construct an explanation for real-world phenomena.
- Apply scientific reasoning to show why the data or evidence is inadequate for the explanation or conclusion.

Materials

None

Advance Preparation

Review student work in Chapters 1, 2, and 3, and be prepared to help students find relevant evidence and ideas in their notebooks.

Phenomena, Data, or Models	Intended Observations	Purpose	Rationale or Notes
Activity 1 Data showing % of Dry Weight of types of atoms making up an evergreen tree shows that C, H, and O atoms make up most of the dry mass of an evergreen tree. Information (based on data) about the mass of C, H, and O atoms is used to calculate the % that each atom contributes to the mass of glucose $C_6H_{12}O_6$.	C, H, and O atoms make up 96% of the dry mass of an evergreen tree. Various minerals, e.g., N, P, Ca, make up about 4% of the dry mass of an evergreen tree. Based on calculations: • C atoms contribute 40% of the total mass of glucose. • H atoms contribute 7% of the total mass of glucose. • O atoms contribute 53% of the total mass of glucose. So CO_2 (which provides C and O atoms) contributes the most to the mass of a growing plant.	Students use experimental data as evidence that minerals contribute very little to the dry mass of an evergreen tree, and perhaps of other trees as well. Calculations based on data show that C and O atoms are responsible for most (93%) of the mass of glucose.	The data will help challenge a common student misconception that most of a tree's mass comes from minerals in the soil. The conclusion that C and H atoms provide most of the mass of glucose will be part of the reasoning in students' explanations of where the mass of Van Helmont's willow tree came from.
Activity 2 Data from Van Helmont's willow tree experiment shows that the increase in the tree's mass cannot be accounted for by the decrease in the mass of the soil.	The willow tree's mass increased by 164.2 lb whereas the decrease in the soil's mass was only 0.1 lb.	Students critique Van Helmont's (invalid) conclusion that the increase in mass came from the water, based on the observation that Van Helmont didn't measure the water and then use what they have learned about photosynthesis, cellulose production, and what they learned in Activity 1 to write a valid explanation for where the increase in mass of the willow tree comes from. Activity 2 can be used as an embedded assessment—rubrics are provided in the online resources.	Van Helmont's data challenges the common misconception that most of the tree's mass comes from the soil. The explanation activity helps students tie together all they have learned about plant growth and enables them to refute a common misconception.

Lesson 3.5—Explaining Where the Mass of Growing Plants Comes From

What do we know and what are we trying to find out?

In Chapter 3, you have investigated how plants grow. You have applied what you learned in Chapter 1 about atom rearrangement during chemical reactions to the growth of plants. In this lesson, you will apply what you learned in Chapter 2 about atom conservation, conservation of total mass, and changes in measured mass to plant growth. Answer the Key Question to the best of your knowledge. Be prepared to share your ideas with the class.

Key Question: Where does the mass of growing plants come from?

The mass of growing plants comes from atoms in the plants' environment. In the last lesson, we showed how C atoms from CO_2 in the environment can become part of cellulose polymers making up plant body structures.

Teacher Talk and Actions

What do we know and what are we trying to find out? (3 min)

Whole class:

Read text under "What do we know and what are we trying to find out?" aloud to the class.

Key Question (3 min)

Key Question: **Where does the mass of growing plants come from?**

Briefly discuss the Key Question. As students share their ideas, ask them to give reasons. Note how many students invoke/agree with general principles from Chapter 2 (e.g., more mass means more atoms) and how many invoke/agree with specific information about plants from Lesson 3.3 (e.g., some of a plant's mass comes from C atoms from CO_2).

Activity 1: Atoms That Contribute Mass to Plant Body Structures

You learned in earlier lessons that plant body structures are made up mainly of carbohydrate polymers such as cellulose and starch and that cellulose and starch are made up of C, H, and O atoms. So does that mean that the mass of a plant is mainly due to the C, H, and O atoms that make up its body structures? If so, do C, H, and O atoms contribute equal amounts to the mass of plants? In this activity, you will examine data to answer these questions.

Procedures and Questions

1. Examine the information in Table 3.8 and answer the questions that follow.

Table 3.8. Percentage of Dry Weight by Type of Atom for an Evergreen Tree

Atom Name(s)	Symbol(s)	% of Dry Weight
Carbon, hydrogen, oxygen (combined)	C, H, O	96.0
Nitrogen	N	1.3–3.5
Potassium	K	0.7–2.5
Calcium	Ca	0.3–1.0
Phosphorus	P	0.2–0.6
Magnesium	Mg	0.1–0.3
Sulfur	S	0.1–0.2

a. According to Table 3.8, which atoms make up 96% of an evergreen tree's dry mass?

Carbon, hydrogen, and oxygen atoms

b. Based on the data in Table 3.8, how does the amount of minerals (N, K, Ca, P, Mg, S) compare with the amount of C, H, and O atoms that make up an evergreen tree?

It's much smaller.

c. How much do you think minerals contribute to the tree's increase in mass as it grows? Why?

They probably don't contribute much to the mass because there are far fewer of those atoms making up plants than there are C, H, and O atoms.

Teacher Talk and Actions	
Activity 1	**(20 min)**
Whole class: Lead a discussion of Step 1 by asking some students to share their answers. Encourage students to challenge each other's ideas when they think the idea is inconsistent with the evidence they have seen.	

d. You have probably heard that plants need sunlight in order to grow. This is true; however, sunlight is not made up of atoms. If sunlight isn't made up of atoms, would it contribute any mass to a growing plant? Why or why not?

Sunlight would not contribute any mass to a growing plant because it is not made up of atoms and so it doesn't have mass.

2. In Lesson 3.3, you determined that the C and O atoms that make up glucose come from CO_2 and that the H atoms come from H_2O. The chemical formula for glucose, $C_6H_{12}O_6$, tells you that glucose is made up of 6 C atoms, 6 O atoms, and 12 H atoms. Which atoms contribute most of the mass of glucose? If all types of atoms had the same mass, then you could figure it out by counting atoms. However, you already know that different types of atoms have different masses. So to answer the question, you need to find out the mass of each type of atom.

   a. Look back at Table 2.3 (p. 78) to find the mass of H and O atoms, and record the information in Table 3.9 below. The mass of C has already been filled in for you.

   b. Using the first row of Table 3.9 as a model, fill in the missing information for H and O, and then calculate the percentage of the total mass of glucose that comes from H and O atoms.

Table 3.9. Composition of Glucose

Type of Atom in Glucose	Mass of Each Type of Atom (amu)	Number of Each Type of Atom in Glucose	Mass (amu) of Atoms of Each Type in Glucose	% of Total Mass of Glucose
Carbon, C	*12 amu*	*6 atoms*	*12 amu/atom x 6 atoms = 72 amu*	*72 amu/ 180 amu = 40%*
Hydrogen, H	*1 amu*	*12 atoms*	*1 amu/atom x 12 atoms = 12 amu*	*12 amu/ 180 amu = 7%*
Oxygen, O	*16 amu*	*6 atoms*	*16 amu/atom x 6 atoms = 96 amu*	*96 amu/ 180 amu = 53%*
			Total mass = 180 amu	100%

3. Based on Tables 3.8 and 3.9, how much of an evergreen tree's dry weight is due to C and O atoms? Some? Most? Almost all? Why do you think so?

Almost all, because C and O atoms contribute 93% (40% + 53%) of the mass of glucose and a tree is mostly cellulose made from glucose monomers.

Teacher Talk and Actions

Activity 1 (cont)

Step 1d: Although the unit does not fully address energy, students are asked to reason that sunlight doesn't contribute to the mass because it isn't made up of atoms. By focusing on atoms, students can draw on knowledge they have (atoms have mass) rather than knowledge they don't have (what energy is).

Let students know that later on in their study of biology, they will learn about the role of energy in chemical reactions, which will help them understand why sunlight is needed even though it doesn't supply any atoms.

Step 2: Completing the table demonstrates quantitatively that most of the mass of glucose comes from C and O atoms.

Step 3: This is intended to bring home the point that very little of the mass of a tree comes from minerals in the soil. In this step, students can estimate the % of a plant's mass is due to C and O atoms (which come from CO_2). A quantitative answer requires that students integrate the information in Table 3.9 with the information in Table 3.8. Specifically, if C, H, and O atoms account for 96% of a plant's dry mass (mostly cellulose made from glucose) and C and O atoms account for 93% of the mass of glucose, then the % of a plant's dry mass that is due to C and O atoms is about 89% (96% x 93%). Students with good math skills may enjoy coming up with a quantitative answer. For others, it's OK for them to appreciate qualitatively that almost all of the mass is due to C and O atoms, which come from CO_2.

Activity 2: Analyzing, Critiquing, and Revising Explanations

Science is a way of explaining the natural world. Sooner or later, the validity of scientific conclusions is settled by referring to observations of phenomena. Hence, scientists concentrate on getting accurate data. As new observations are made and as technologies make possible more accurate measurements, new evidence can cause scientists to question their current explanations. This activity provides an opportunity for you to examine how an explanation changed over time.

Procedures and Questions

1. The following paper describes an experiment carried out by a scientist named J. B. Van Helmont that was published nearly four centuries ago (1662).
   a. Read the paper on this page and the next, underline Dr. Van Helmont's claim, and circle his summary of his evidence.

When a Willow Tree Grows, the Increase in Mass Comes Mainly From Water

Introduction

People have long wondered where the mass of a large tree comes from. Some have argued that the mass comes mainly from the soil. The experiment described below was designed to find an answer to the question of where the mass comes from when a giant tree grows from a tiny seedling.

Methods

I took 200 pounds of soil dried in an oven, moistened it with rainwater, and planted a 5 pound willow tree in it. Whenever the tree needed watering, I watered it with either rainwater or distilled water. After five years, I took out the tree, shook all of the soil out of its roots, and weighed it. I also dried all the soil the tree had grown in and weighed the soil again.

Teacher Talk and Actions	
Activity 2	**(25 mins)**

Read the text under Activity 2 aloud to your students.

Step 1a. Have students read the description of Van Helmont's experiment, underline his claim, and circle his summary of the evidence. Circulate to be sure each student

• underlines the title of the paper as his claim: The increase in mass of a willow tree when it grows comes from water, and

• circles the statement in the discussion, "and found that the increase in the mass of the tree was far more than the decrease in mass of the soil" as the summary of his evidence.

Results

After five years, the willow tree weighed 169 pounds and about three ounces and the dried soil weighed only a couple of ounces less than its former weight. The results are shown in Table 1.

Table 1. Results of Dr. Van Helmont's willow tree experiment

			Soil		Tree
Day 1		=	 200.0 lbs	+	 5.0 lbs
Five years later		=	 199.9 lbs	+	 169.2 lbs
Difference			-0.1 lbs		+164.2 lbs

Discussion and Conclusion

My experiment was designed to determine the source of the mass of a growing tree. I compared the increase in mass of the tree with the decrease in mass of the soil and found that the increase in the mass of the tree was far more than the decrease in mass of the soil. Since the only thing I added to the tree was water, the increase in mass must have come from the water.

b. Does Van Helmont's evidence support his conclusion? Why or why not? Does his explanation meet the elements of the Explanation Quality Criteria?

No, Van Helmont's evidence does NOT support his conclusion. He didn't collect any data about water, so he can't conclude that water was the source of the increased mass of the willow tree. Van Helmont can conclude from his data that the soil is not the source of the increased mass of the willow tree, but he has no data to provide evidence of where the increased mass comes from.

Teacher Talk and Actions
Activity 2 (cont)
Step 1b. Have students work individually to decide if Van Helmont's data supports his claim. Check to see if students are providing evidence they know about that is inconsistent with Van Helmont's claim. It's fine if they cite data from the unit, but it is not necessary. If they do, ask if Van Helmont would have had access to this data.

c. Write a better explanation for where most of the mass of a willow tree comes from. You may work with your partner or group to fill in the table below. However, once the table is complete, you should work independently to write your own explanation in the space provided on the next page.

Question	Where does most of the mass of a willow tree come from?
Answer	*Most of the mass comes from carbon dioxide (CO_2) in the air.*
Science Ideas	*Carbohydrates are the main polymers making up plant body structures. Carbohydrate polymers are molecules made of glucose monomers. Different carbohydrate polymers have different properties because they are made of different numbers and arrangements of glucose monomers (Science Idea #11).* *Plants use carbon dioxide and water molecules in their environment to make glucose and oxygen molecules. Atoms are rearranged during this chemical reaction (Science Idea #12).* *To build body structures for growth and repair, plants use glucose monomers to make carbohydrate polymers and water molecules. Atoms are rearranged during this chemical reaction (Science Idea #13).* *When plants grow or repair, they increase in mass. Atoms are conserved when plants grow: The increase in measured mass comes from the incorporation of atoms from molecules that were originally outside of the plants' bodies (Science Idea #14).*
Evidence	• *Plants' body structures are mostly made up of carbohydrates (Table 3.2, p. 105).* • *96% of plants' dry mass is due to C, H, and O atoms (Table 3.8, p. 134).* • *C and O atoms contribute 93% to the mass of glucose ($C_6H_{12}O_6$) (Table 3.9, p. 135).* • *Algae grown with labeled $^{14}CO_2$ make ^{14}C-glucose (Table 3.6, Experiment 1, p. 119).* • *Algae grown with ^{18}O-labeled H_2O produce $^{18}O_2$, not $C_6H_{12}{}^{18}O_6$ (Table 3.7, Experiment 2, p. 119).* • *Mouse-ear cress plants produce ^{14}C-cellulose from ^{14}C-glucose (Table 1 in Experiment, p. 127).*
Models	*Modeling activities in Lessons 3.3 and 3.4 showed us that* • *it is possible to make $C_6H_{12}O_6$ and O_2 by rearranging atoms of CO_2 and H_2O molecules, and* • *it is possible to make cellulose and water molecules by rearranging atoms of glucose molecules.*

Teacher Talk and Actions
Activity 2 (cont)
Step 1c. Use this as an assessment. Have students work individually to write a better explanation than Van Helmont's. Be sure students are using ideas about atoms and molecules (i.e., atom rearrangement, conservation, and changes in measured mass). Students can work in pairs or small groups to complete the table. Circulate as students are working to see if they are trying to find as much evidence as possible. Ask, "I see a group that has identified two pieces of evidence. Has any group identified more than two? More than three?" and so on. Once students are pretty sure they have all the evidence they need, have each student work independently on his or her explanation.

Explanation:

The mass of the willow tree comes mainly from carbon dioxide in the air outside the plant. Plant body structures are mostly made up of carbohydrates (Table 3.2, p. 105). Mouse-ear cress plants produce cellulose from glucose (Table 1, p. 127) and Science Idea #13 states that plants can make carbohydrate polymers from glucose monomers, so the willow tree can probably make cellulose from glucose. Science Idea #12 states that plants make glucose from CO_2 and H_2O and data from the labeling experiments (Tables 3.6 and 3.7, p. 119) showed that the C and O atoms of glucose come from CO_2. Our modeling activities showed us that it is possible to make $C_6H_{12}O_6 + O_2$ by rearranging atoms from $CO_2 + H_2O$ and that it is possible to make cellulose (and H_2O) by rearranging atoms from $C_6H_{12}O_6$. So I think the willow tree can take CO_2 from the air and H_2O from the soil and rearrange atoms to make glucose (and O_2) and then react glucose monomers to make cellulose (and H_2O) by rearranging atoms from carbon dioxide and water. Data in Table 3.8 (p. 134) shows that 96% of plants' dry mass is due to C, H, and O atoms. And data in Table 3.9 (p. 135) shows that C and O atoms contribute 93% of the mass of glucose.

So CO_2 supplies the atoms that make up most of the mass of the willow tree. Science Idea #14 states that the increase in mass of a growing plant is due to the incorporation of atoms that were originally outside the plant. So I think the willow tree incorporates carbon dioxide into glucose, then glucose into cellulose when a willow tree grows.

Teacher Talk and Actions
Activity 2 (cont)
(Intentionally left blank)

Pulling It Together

1. Now that you have explained where the added mass of Van Helmont's willow tree came from, how would you answer the Key Question, **Where does the mass of growing plants come from?**

   The increased mass of growing plants comes mainly from carbon dioxide in the air. Carbon dioxide provides the carbon and oxygen atoms plants use to make glucose. Plants then use the glucose to make cellulose, which most plant body structures are made up of.

2. Can you think of other examples of a scientific explanation that has changed over time? What new evidence was found to support the revised explanation?

   From thinking that diseases result from supernatural causes to the germ theory of disease: After collecting microorganisms from a diseased cow, Pasteur showed that injecting some of the microbes into a healthy cow caused the cow to become sick whereas injecting some heat-killed microbes into a cow did not.

   From thinking that the surface of the Earth is solid and unchanging to the theory of plate tectonics: Satellites can detect that continents are moving at about 2 mm per year OR examination of world maps showed that the continents of Africa and South America fit together as if they were once a single continent.

   From thinking that the Earth is flat to thinking that the Earth is a sphere: even before spaceships and satellites could photograph the Earth from outer space, people noticed how a ship would disappear over the horizon, as if it was sinking. The ship was actually "sinking" around the curve of the Earth.

   From thinking that the sun and planets revolve around the Earth to thinking that the Earth and other planets revolve around the sun.

   From thinking that eating cholesterol is harmful to thinking it may not be.

   From thinking that particular mental disorders were due to character flaws to recognizing that there is an underlying chemical basis for them.

3. Are there other situations outside of science where using evidence-based explanations is important? If so, what are some examples?

   In court cases, lawyers for the prosecution and defense have to use evidence-based explanations to make a convincing case to the jury. Even in our everyday lives, we need to use evidence to make good decisions about many things, for example, which new car or appliance is the most energy-efficient at the price we can afford, what foods are the most nutritious to eat, and who to vote for based on their record and not just on their ads.

Teacher Talk and Actions
Pulling It Together **(10 mins)**
For Pulling It Together Question 1, students should respond to the Key Question by stating what plant body structures are made of at the molecular level (cellulose polymers), where the cellulose comes from (plants make it from glucose), and where the atoms of glucose ($C_6H_{12}O_6$) come from (CO_2). Because both C and O atoms are much heavier than H atoms, the C and O atoms contribute much more mass to glucose than the H atoms. Pulling It Together Questions 2 and 3 are designed to find out in a general way the extent to which students understand how scientific knowledge changes over time as a result of new evidence. For Question 1, an initial whole class brainstorming session focused on identifying developments in microscopes, telescopes, imaging technologies, and biotechnology might be a useful strategy for stimulating their thinking.
Closure and Link **(3 min)**
In Chapter 3, we applied our knowledge of chemical reactions to explain growth and repair in plants. In the next chapter, you will have the chance to learn about growth and repair in animals.

Toward High School Biology

Chapter 4

Teacher Edition

Chapter 4 Overview

This chapter focuses on explaining the growth and repair of animal body structures such as muscles, tendons, skin, fur, scales, and feathers. The body structures of animals are made largely of polymers, especially proteins. Students were first introduced to the nylon synthesis reaction in Chapter 1 so that they would become familiar with how polymers are made from monomers. They also saw that glucose monomers could make cellulose polymers in Chapter 3. Animals grow by making their body proteins from the proteins in their diet: Protein polymers in their food are first broken down to amino acid monomers in the digestive system, and then the amino acids are transported by the blood to body cells, where they are used to build their own body proteins. Both protein digestion and protein synthesis occur through chemical reactions, in which atoms are rearranged but not created or destroyed. Animal growth does not violate the conservation of matter; the atoms that contribute to the increased mass come from the food animals take in from their environment.

Lesson 4.1 presents students with several examples of animal growth and repair (a puppy growing, a human girl growing, a lobster growing, and a lizard replacing its lost tail), gives students data on the mass being gained in each case, and asks them to think about what new materials are being made. It also looks at protein, fat, and carbohydrate molecules and shows that protein makes up the largest percentage of an animal's mass (as opposed to carbohydrate or fat) and that muscle makes up the largest portion of the human body by mass (as opposed to skeleton, fat, skin, etc.), thereby providing a rationale for focusing on muscle protein to explain growth in subsequent lessons.

Lesson 4.2 presents data showing that different proteins have different properties (water solubility, fibrous/nonfibrous) and have different sequences of amino acid monomers. Students are first shown an egg-eating snake that eats only eggs, made up mostly of the protein ovalbumin. The snake sheds its skin (mostly keratin) regularly and clearly needs to build keratin. Based on the evidence that proteins making up an animal's body structures are not the same substances as the proteins in its diet, students should reason that chemical reactions must take place for an animal to grow and build its body structures. To understand why different proteins have different properties, students examine the basic structure of protein polymers, the amino acid monomers they are made from, and partial amino acid sequences of egg-white ovalbumin and actin to see that they differ.

Lesson 4.3 uses radioactive labeling, which students were introduced to in Chapter 3, to track atoms in an experiment that shows that herring fish digest their food (brine shrimp) and some of the atoms from the food end up as part of the fish's growing bodies. Students then model the reactions that happen inside the fish. The first type of reaction involves the breakdown of proteins into amino acid monomers during digestion, and the second type of reaction involves the building of new proteins from amino acid monomers in a different order. Students start with one carbon atom "labeled" (with a sticker) and then look to see where it ends up at the end of both reactions (in the new protein in the fish's bodies). The fact that the labeled carbon atom starts in one substance and is later found in another is evidence that atom rearrangement, and hence chemical reactions, occur.

Lesson 4.4 functions as an embedded assessment that requires students to synthesize ideas and practices from Chapters 1, 2, and 3. Students analyze the authors' explanation in a simplified version of a published science paper that uses data from a radioactive labeling experiment to show that elderly men fed a complete diet plus an amino acid supplement make more muscle protein than a control group fed only the complete diet. They then write an explanation to refute the claim that "chemical reactions only happen in labs, not in people or animals."

Lesson 4.5 gives students an opportunity to apply everything they have learned about plant and animal growth to a new type of living thing: mushrooms. It is intended to help students generalize the ideas that they have learned to all organisms. Students read some basic information about mushrooms and are asked to reason about how they might grow.

Chapter 4 Background Knowledge for Teachers

Chapter 4 presents a basic story about chemical reactions that contribute to the growth of animals: Animals eat food that contains proteins. These proteins are broken down via chemical reactions to their amino acid monomers. Through other chemical reactions, animal bodies use these amino acid monomers to build whatever proteins they need for growth and repair. The atoms from food that were once outside the body are now part of the body structures (as part of protein molecules). The addition of atoms means the body has more mass, and it has grown.

The unit limits its treatment of proteins to structural proteins (not enzymes, membrane proteins, antibodies, or carriers such as hemoglobin). It focuses on proteins inside cells, such as muscle actin and myosin, and the extracellular protein collagen used to build connective tissue. It also focuses on protein polymers animals use to build external body structures, for example, the keratin that animals make for hair, skin, nails, horns, claws, and scales.

Obviously, for an animal to grow, it has to build new cells and all their molecular components. For example, organisms need to build lipids for cell membranes and replicate DNA to provide genetic instructions for each new cell. The unit does not address lipids or DNA or cell division. In fact, the unit does not address the cellular level at all—it simply shows that protein molecules build structures. This background information will supply you with knowledge of how cells relate to growth, but it is only for your information. You might also want to use it (sparingly) if a few advanced students ask.

Proteins

The work of the cell is carried out by the many different types of molecules it assembles, mostly protein polymers. Protein polymers are long chains of amino acid monomers that are folded, twisted, or braided into complex structures. Cells make protein polymers from 20 different kinds of amino acid monomers. Cells make thousands of different proteins that vary in length from about 40 to over 33,000 amino acids and that vary in amino acid composition and sequence. The function of each protein depends on its specific sequence of amino acids and the shape the chain takes as a consequence of attractions between the chain's parts. Some proteins assist in replicating genetic information, repairing cell structures, helping other molecules get in or out of the cell, and generally in catalyzing and regulating molecular interactions. In specialized cells, other protein molecules may carry oxygen, effect contraction, respond to outside stimuli, or provide material for hair, nails, and other body structures. In still other cells, proteins may be exported from the cells that make them to serve as hormones, antibodies, or digestive enzymes.

Amino acid monomers have a common structure: an amino ($-NH_2$) group, a carboxyl (-COOH) group, a hydrogen atom (-H), and an "R" group on the same carbon atom. (See next page for the common structure of an amino acid.)

While the amino and carboxyl groups are the same in all amino acids, it is the R groups that distinguish one amino acid from another. The R groups vary in size, charge, and chemical reactivity, making possible a wide variety of proteins with diverse functions.

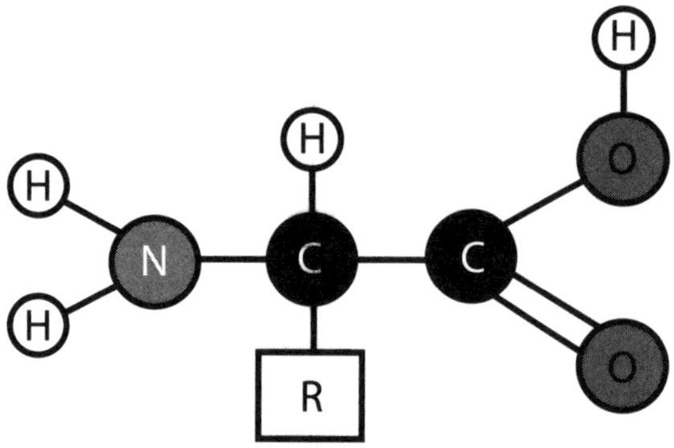

Amino acids form protein polymers through chemical reactions that remove an H from the NH_2 (amino) group of one monomer and an OH from the COOH (carboxyl) group of another monomer to form an OC—NH linkage (peptide bond) and a water molecule.

Humans can synthesize approximately half of the 20 amino acids from glucose and nitrogen obtained from other amino acids. The other half cannot be synthesized from glucose and must be consumed in the diet. For this reason, they are called "essential" amino acids.

Plants can make all 20 amino acids and all other biomolecules from glucose and usable sources of elements such as N, P, K, S, Ca, and Mg. However, some of the amino acids are not in sufficient supply to satisfy the needs of most animals. Some herbivores rely on the growth of bacteria on the plant material in their stomach (a fermenting chamber) to produce amino acids.

Carbohydrates

In Chapter 4, carbohydrates are shown as a type of molecule that is a component of most animal bodies. But since proteins account for most of the mass of animal body structures, this chapter uses proteins to explain animal growth.

However, the growth of mushrooms is explored in Chapter 4 in a capstone lesson that incorporates ideas about growth from Chapters 3 and 4. Mushroom growth—that is, the growth of an organism that is neither plant nor animal—involves the production of chitin, a carbohydrate polymer that also contains nitrogen atoms. The structure of chitin is omitted to avoid unnecessary detail.

Lipids

Lipids provide the structural foundation for biological membranes, and this is their most important role. Because the unit does not focus on cells, and because the structures and biosynthesis of lipids are more complex than those of proteins, the unit does not include lessons on the structure and synthesis of lipids.

Nucleic Acids

Nucleic acids, including DNA and RNA, are essential biopolymers for all living organisms. The genetic information encoded in DNA molecules directs the sequencing of amino acids in proteins. DNA molecules are the largest known individual biomolecule: A single human chromosome can contain more than 200 million nucleotide pairs (in contrast to ~1,000 glucose monomers in starch and ~33,000 amino acids in the largest protein). While the unit does not focus on DNA or its role in directing the synthesis of proteins, the information it provides about the chemistry of protein polymer formation will be useful when students study the relationship between DNA, proteins, and traits in high school biology.

Digestion

When animals digest proteins, the proteins react with water to form amino acid monomers. Yet despite the fact that proteins and water molecules are abundant in the stomach, protein digestion does not occur without enzymes. (Protein breakdown does occur if proteins are cooked for many hours, but living organisms wouldn't survive this treatment.) What students are not shown is that there is a variety of digestive enzymes (pepsin, trypsin) that serve as catalysts. Stomach acid (hydrochloric acid, or HCl) lowers the pH of the stomach environment so that the enzymes will work well. Some enzymes can operate at the higher pH of the small intestine. Because students are just starting to solidify their ideas about chemical reactions, ideas about enzymes would overly complicate the storyline.

Energy

Carbohydrates are both building blocks for complex molecules and also fuels to satisfy the energy needs of cells. The unit does not discuss the metabolism of carbohydrates to release energy for the cell. We made this decision for both practical and conceptual reasons. The time that school districts were able to devote to chemical reactions was insufficient to develop coherent stories of both matter transformation and energy transformation. Our observations indicated that students starting the unit lacked mental models of both matter transformation and energy transformation, and six weeks was not enough time to help students develop and use both. Because matter transformation could be made concrete with physical models, we opted to start there.

It is important to note that we initially tried to include a partial energy story—the rule that "atoms do not turn into energy, and energy does not turn into atoms"—but found it to be unsatisfying to most students. While a few students were able to cite the rule when challenging classmates' explanations that involved creating matter from energy, the rule didn't appear to convince their classmates. We think that students will be less likely to invoke matter-from-energy explanations of chemical reactions if they have a strong mental model of atom rearrangement and conservation.

Lesson Guide
Chapter 4, Lesson 4.1
The "Stuff" That Makes Up Animals

Focus: This lesson focuses on the composition of animal bodies, emphasizing that they are made up of matter (atoms and molecules) and that protein molecules are an especially important contributor.

Key Question: What substances do animals need in order to grow and repair?

Target Ideas Addressed

- Animals gain mass when they grow.
- Animal bodies build many different body structures as they grow and repair.
- These body structures are made up largely of protein molecules.

Science Practice(s) Addressed

- Analyze and interpret data to provide evidence for phenomena.
- Apply scientific ideas, principles, or evidence to construct an explanation for real-world phenomena.

Materials

Activity 1: Projector, Videos: *Dog Growth* and *Human Girl Growth*; Slide shows: *Lobster Growth* and *Repair in a Lizard*; Basic calculators for students (optional)

Activity 2: Projector, Video: *Chicken Muscle and Tendon,* Slide show: *Interior Structures of Animals*

Advance Preparation

Locate videos and slides on the THSB website and set up the projector.

None of the videos or slide shows will have audio, so review them and use the teacher facilitation pages to prepare to narrate them for students.

Phenomena, Data, or Models	Intended Observations	Purpose	Rationale or Notes
Activity 1 Videos and photos of growth and repair: dog, human, lobster, and lizard	Animals (including humans) grow and repair body structures. They build new materials when they grow and repair body structures.	Students begin to think about animals (including humans) as being made up of matter and that, because they are made up of matter, the ideas in the chemistry lessons will apply to animals.	The animals were selected to be both familiar to students and taxonomically diverse.
Activity 1 Data on mass gained during growth: dog, human, lobster, lizard	Animals gain mass when they grow and when they repair their bodies.	Students examine data about growth in different animals to ensure that students associate growth with a gain in mass (as opposed to, e.g., just getting bigger, maturing, or aging).	Tail regeneration in lizards was selected as an example of repair because smaller wounds do not illustrate mass change as clearly and visibly.
Activity 2 Diagrams of musculature of animals	Muscles and tendons make up a very large portion of the interior of most animal bodies.	Students use diagrams to begin thinking beyond the surface level in terms of the materials the body makes as it grows.	
Activity 2 Data on protein, fat, and carbohydrate content of a variety of animal bodies	Animal bodies usually contain more protein molecules than fat or carbohydrate molecules.	Students examine data about the content of animals' bodies to help them focus on protein molecules as the most important contributor to animal growth. (Proteins will remain the focus for the rest of Chapter 4.)	The broader term *lipids* is avoided (using the more specific term *fat* instead) because the term *fat* is more familiar to students, and lipids are not the focus of either Chapter 3 or 4.
Activity 2 Data on the proportion of body weight accounted for by different materials (muscles, skin, bones, etc.) making up the human body	Muscle makes up the largest portion of the human body by weight.	Students use data on the materials in the human body to determine that muscle, and therefore protein molecules, are central to animal growth.	

Lesson 4.1—The "Stuff" That Makes Up Animals

What do we know and what are we trying to find out?

In the previous chapter you observed and modeled chemical reactions that contribute to plant growth. You figured out that plants grow by building carbohydrate polymers from glucose monomers they make from carbon dioxide and water molecules in their environment. You started by observing what body parts are getting bigger and heavier as plants grow and then looked at the molecules those body parts are made up of and the chemical reactions that produce those molecules. We will take a similar approach to understanding animal growth.

Answer the Key Question to the best of your knowledge. Be prepared to share your ideas with the class.

Key Question: What substances do animals need in order to grow and repair?

Animals are made of many different things, like skin, fur, scales, shells, nails, hair, muscles, bones, and organs. All of those things are matter, and so they must be made up of atoms and molecules.

Teacher Talk and Actions
What do we know and what are we trying to find out? (3 min)
Whole class: Read text under "What do we know and what are we trying to find out?" aloud to the class.
Key Question (5 min)
Individuals: Students respond to the Key Question: **What substances do animals need in order to grow and repair?** Whole class: Lead discussion. Invite student volunteers to share ideas about the Key Question. The goal is primarily eliciting and probing student ideas here. When students share their answers, do not indicate if their answers are right or wrong. However, do not let students assume by your silence that they were correct. Make it clear that students are expressing ideas, but those ideas may change based on what they will see in the lessons. (Remind students of this regularly.) Challenge any ideas that are inconsistent with learning from previous lessons.

Activity 1: Observing Animal Growth

Different animals have different body structures. Some animals have wings, and others have arms. Some animals have feathers, and others have fur. Some animals have an internal skeleton, and others have an external skeleton. In this activity, you will observe what happens to body structures as animals grow.

Procedures and Questions

1. Observe the videos and photos of animal growth and repair. As you answer the questions below, think about the body structures animals have that you can see and also the body structures you can't see.

   a. What body structures are getting bigger as the **puppy** grows?

   Fur, bones, skin, muscles, claws, teeth, heart, stomach, brain, paws, tail

   b. What body structures are getting bigger as the **human girl** grows?

   Skin, hair, bones, muscles, brain, heart, lungs, stomach, hands, feet, ears

   c. What body structures are getting bigger as the **lobster** grows?

   Shell, claws/pinchers, antennae, legs, tail, muscles (Some students might call this "meat.")

   d. What body structures are getting bigger or being rebuilt as the **lizard** regrows its lost tail?

   Scales, muscles, bones

   e. Do you think the repair of the lizard's lost tail is similar to growth? Why or why not?

   Yes, because something gets bigger or appears that wasn't there before.

   f. Where do you think the "stuff" comes from to make all of these body structures?

   Food

Teacher Talk and Actions

Activity 1 (20 min)

Read the text underneath Activity 1 aloud to the class.

Play the dog growth video for students. (Feel free to pause or reshow parts of it.) Ask students if they remember seeing this video at the beginning of the unit. Do they have new ideas or observations now?

Have students write down as many structures as they can think of that are getting bigger when the puppy grows.

Play the human growth video for students. Have students write down as many structures as they can think of that are getting bigger.

Project the lobster growth slide for students. Have students write down as many structures as they can think of that are getting bigger.

Project the lizard tail regeneration slide for students. Have students write down as many structures as they can think of that are getting bigger.

Science Note to Teacher:

Many lizards purposefully lose their tails when a predator attacks. The predator gets the tail, and the lizard escapes.

Give students time to answer Questions 1e and 1f. Ask a few students to share their answers and have a brief discussion.

Question 1e: Do you think the repair of the lizard's lost tail is similar to growth? Why or why not?

Question 1f: Where do you think the "stuff" comes from to make all of these body structures?

2. You have just observed that various animal body structures get bigger as they grow, but how do we know that animals are actually adding new matter to their bodies? To answer this question, we will observe what happens to the mass as animals grow and repair. Examine the data in Table 4.1 and answer the questions that follow.

Table 4.1. Typical Change in Mass (by Weight in Pounds or Grams) of Some Growing Animals, Over Different Time Periods

Animal	Initial Weight (Age)	Final Weight (Age)	Change in Weight	Increase or Decrease?
German shepherd dog (male)	20 pounds (at 2 months)	51 pounds (at 9 months)	31 pounds	+
Human (female)	7.7 pounds (at birth)	64 pounds (at 10 years)	56.3 pounds	+
Spiny lobster	3 pounds (at 11 months)	4 pounds (at 2 years)	1 pound	+
Lizard	0.62 grams (at time of tail loss)	1.72 grams (when new tail was fully grown)	1.10 grams	+

a. Using the data in Table 4.1, calculate the change in weight for each animal (you can use a calculator). Put a plus sign (+) next to any change in mass that was an increase. Put a minus sign (–) next to any change in mass that was a decrease.

b. Based on these data, what happens over time to the mass of an animal when it grows or repairs its body?

The mass of an animal increases when it grows or repairs, because all of the examples in the table showed an increase in weight over time (we haven't looked at all animals, but based on the ones we've seen, this seems to be the case). The lizard re-growing its tail didn't gain as much mass as the others, but it increased a lot relative to its initial weight. (Note that the lizard's final weight is about 2.8 times its initial weight, whereas the dog's final weight is about 2.6 times its initial weight.)

Teacher Talk and Actions

Activity 1 (cont)

Have students read Table 4.1 and let them use calculators to determine the change in weight for each animal, then mark a + or – to indicate an increase or decrease.

Note: Each of the animals was tracked over a different time span (due to different life spans and growth rates), so focus students on "initial" and "final" if they are getting confused by the times in parentheses.

Make sure that all students understand that animals increase in mass when they grow before you move on from this activity.

Give students time to answer Question 2b.

Some students might be in the habit of saying, "it stays the same," when asked questions about mass or the number of atoms.

They are asserting an understanding of conservation of mass, but make sure they are paying attention to the question being asked. When an animal grows and gains weight, the measured mass of the animal increases.

Push students to think about the implications of an increase in weight:

If each of the four types of animals increased in weight, what is happening to the amount of matter as they grow? What must be happening to the number of atoms making up the animals' body structures? Where do you think these atoms could have come from?

Activity 2: The Substances That Make Up Animals

As you have seen, animal bodies are very complex. They have a wide variety of body structures—but what makes up those structures? For example, you may have heard that 50%–60% of a human's weight is water. So what keeps us from being puddles on the ground? Animals, including humans, are made up of many other substances besides water, and these substances make up structures that give animals their shape. Aside from water, most of an animal's body is made up of three major types of substances: *proteins, fats*, and *carbohydrates*. You learned in Chapter 3 that although plants are made up of the same three major types of substances, plant body structures are made mostly of carbohydrates. Is this also true for animals? In this activity, you will find out how much of each type of substance makes up the bodies of various animals.

Procedures and Questions

1. Observe the photos and videos of the interior body structures of animals. Think about what they might be made of.

2. With your group, examine Tables 4.2 and 4.3 (p. 146) and answer the questions that follow.

   a. On Table 4.2, **circle** the type of molecule that is present in the highest amount in all the animals.

Table 4.2. Relative Mass (in Grams per 100 Grams) of Protein Molecules, Fat Molecules, and Carbohydrate Molecules in Some Animal Bodies

Animal	Protein (per 100 g)	Fat (per 100 g)	Carbohydrate (per 100 g)
Anchovy fish	65.3	7.1	4.4
Brine shrimp	58.4	4.0	29.3
Cow	60.0	8.9	15.1
Chicken	42.3	37.8	10.5
Crayfish	47.5	3.4	29.3
Hamster	49.8	34.6	8.1
Herring fish	72.7	8.5	0.8
Human	48.0	38.7	1.5
Locust	22.1	3.0	4.3
Squid	74.8	8.8	4.9
Worm	56.4	7.8	19.6

Teacher Talk and Actions
Activity 2 **(10 min)**
Read the text under Activity 2 aloud to the class. Project the diagrams of the interior structures of animals. These images focus on muscles and tendons, which will be central throughout Chapter 4. Give students some time (3–5 minutes) to study Table 4.2. Note: Students may notice that crayfish and brine shrimp have relatively high amounts of carbohydrate. It's the cellulose and chitin that makes up their exoskeletons. If students ask, you could ask students what they think might be different about the body structures of brine shrimp and crayfish compared with the other animals listed in the table. Students should notice that both have exoskeletons. However, a locust has an exoskeleton but doesn't have high carbohydrate amount and a worm doesn't have an exoskeleton and yet still has higher carbohydrate amount. So there must be other factors. *This is a digresion and should only be addressed if students ask.*

Table 4.3. Average Weight of Body Materials in a 154-Pound Human

Material	% of Body Weight
Muscle	41.5
Fat	18.0
Skeleton	15.8
Blood	8.0
Skin	6.9
All the rest (brain, liver, kidneys, etc.)	9.8

b. In the coming lessons, we will focus on muscle to understand animal growth. Based on the data in Table 4.3, why do you think it might be a good idea to focus on muscle?

The table shows that muscle makes up more of the human body by weight than any other material. That means it's likely that when any animal grows and increases in mass, most of it is muscle. So studying muscle will help us figure out a lot about animal growth.

c. Based on the data in Tables 4.2 and 4.3, what type of substance do you think animals will need to make the most of in order to build body structures for growth? Why do you think so?

Animals will need to make mostly protein to grow, because there are more proteins than carbohydrates or fats in most animal bodies. Also, muscle, which is mostly proteins, makes up more of a human's weight than any other material.

Teacher Talk and Actions
Activity 2 (cont)
Give students some time to study Table 4.3 and talk about it as a group. Encourage students to work with their group to answer Questions 2b and 2c. Look to see if students generalize from the human example to all animals. If they do, look to see if they write qualifying words such as "maybe" in recognition of not having direct evidence. Students are encouraged to be clear about what they have evidence for and what they do not. Students might not be comfortable with being uncertain, but encourage them that they don't need to get everything "right" immediately.

Pulling It Together

Work on your own to answer these questions. Be prepared for a class discussion.

1. Now that you have seen how much mass animals gain when they grow, and the substances that make them up, how would you answer the Key Question, **What substances do animals need in order to grow and repair?** Provide evidence for your answer.

Data in Table 4.2 shows that in addition to water molecules, a wide variety of animals are made up mostly of protein molecules (in comparison to carbohydrate or fat molecules). The animals listed include both animals with bones (internal skeletons) and animals without bones (external skeletons), so I think that in addition to water molecules probably all animals are mostly made up of protein molecules.

2. If mass is increasing as animals get bigger, what does that mean about the number of atoms making up their bodies?

Data in Table 4.1 shows that a growing dog, human, lobster, and lizard increase in mass. If mass increases as animals get bigger, then the number of atoms is probably increasing, because mass is due to atoms (based on Science Idea #8).

3. Think back to the animal body structures that you have observed, such as hair, fur, skin, muscle, tendons, scales, and scar tissue. We just concluded that these body structures are mostly protein. Do you think all proteins are the same substance or different substances? Why or why not?

Muscles and tendons making up chicken meat seem to have different properties. Maybe the proteins making up different body structures are different.

Teacher Talk and Actions
Pulling It Together **(10 min)**
Individuals: Students answer Pulling It Together questions on their own, in detail. Whole class: Lead a class discussion. Invite students to share ideas about Pulling It Together Question 1, which returns to the lesson Key Question. Direct students to look back at their initial ideas about the Key Question and consider how their ideas have changed. Students should have an improved answer for Pulling It Together Question 1, the Key Question. Pulling It Together Question 2 is meant to get students to explicitly connect an increase in mass in a living organism to an increase in atoms. Pulling It Together Question 3 is meant to get students to begin to link differences in properties of materials making up body structures and differences in the molecules that make them up.

Lesson Guide
Chapter 4, Lesson 4.2
Proteins in Animal Bodies and Food

Focus: This lesson introduces students to protein molecules: some of their functions, their basic molecular structure (polymers made up of amino acids), the idea that there are many different proteins (which are different substances—not just "protein"), and the fact that the proteins an animal eats are usually not exactly the same types that make up its body.

Key Question: Are the proteins that animals eat exactly the same as the proteins that make up their own bodies?

Target Ideas Addressed

- Protein molecules are polymers made from amino acid monomers.
- Different proteins have different types, numbers, and arrangements of amino acid monomers, and therefore are different substances with different characteristic properties.

Science Practice(s) Addressed

- Use a model to describe phenomena.
- Apply scientific ideas, principles, or evidence to construct an explanation for real-world phenomena.

Materials

Activity 1: Projector, Slide show: *Nursing Puppies*, Video: *Egg-eating Snake,* Slide show: *Snake Shedding Skin*

Advance Preparation

Locate videos and slides on the THSB website and set up the projector.

None of the videos or slide shows will have audio, so review them and use the teacher facilitation pages to prepare to narrate them for students.

Phenomena, Data, or Models	Intended Observations	Purpose	Rationale or Notes
Activity 1 A nursing puppy drinks milk but needs to build fur, muscle, and other body structures.	The puppy eats only milk (casein) but builds its fur and muscles of keratin and myosin, neither of which are found in milk.	Students are provided with a clear example of an animal that eats only a certain kind of protein but needs to make different proteins from the one it eats. (This example should prompt the idea that chemical reactions must be involved in making new proteins.)	Puppies were chosen because they clearly only drink milk, yet make a multitude of proteins. They provide a clear example of how animals can build different proteins from the ones they eat.
Activity 1 Video of egg-eating snake and photos of snakes shedding their skin	The snake eats only egg protein (ovalbumin) but builds its scales from keratin and its muscles from myosin, neither of which is found in the eggs it eats.	Students are provided with a clear example of an animal that eats only a certain kind of protein but needs to make different proteins from the one it eats. (This example should prompt the idea that chemical reactions must be involved in making new proteins.)	Egg-eating snakes were chosen because their ability to swallow eggs whole and to shed their skin could be engaging to students, and also because these snakes eat only eggs, so they provide an especially clear example of how animals can build different proteins from the ones they eat.
Activity 1 Data on properties (solubility and fiber formation) of several proteins	Different proteins have different properties and, therefore, they are different substances.	Students are shown evidence that different proteins are different substances (and that some are better for building particular animal body structures than others).	Solubility and fiber formation were chosen because students have some experience with solubility from Chapter 1, and both these properties contribute to whether or not a protein would be a good structural protein.
Activity 2 Illustrations of the molecular structure of a small piece of a protein molecule (made from six amino acids) and those six amino acids as free amino acids	Proteins are polymers made from amino acid monomers. Proteins and amino acids are made largely from C, H, O, and N atoms.	Students become familiar with the fact that proteins are polymers made up of amino acid monomers that are made largely from C, H, O, and N atoms as well as with recognizing amino acids within a protein polymer.	These diagrams are flattened and arranged to maximize the visibility of the parts of the structure common to all amino acids: The amino and carboxylic acid groups on the ends. To help students focus on the NH_2 and COOH groups involved in peptide bond formation, the examples students see do not include amino acids with additional amino or carboxylic acids in their side chains (R groups) or amino acids that have sulfur atoms. This keeps the visual pattern as clear as possible.
Activity 2 Data on sequence of chicken egg white and rabbit muscle protein	Different proteins have different amino acid sequences (different arrangements of atoms) and therefore are different subtances.	Students are given evidence that different amino acid sequences are associated with different proteins. Therefore the arrangement of atoms, like in previous examples, is associated with different substances.	Actin and ovalbumin were chosen because they were similar in length, and the students had additional information about both from the previous table.

Lesson 4.2—Proteins in Animal Bodies and Food

What do we know and what are we trying to find out?

In the last lesson, you saw that animal body structures are largely made up of protein. Nutrition labels show the amount of protein in different foods, so we know that we can obtain protein from eating foods with protein, such as meat. Body builders say they need to eat a lot of protein in order to build muscles. Are all proteins the same substance? If different proteins are different substances, what makes them different?

You will learn about the proteins that make up animal bodies, as well as the proteins that animals eat when they eat meat. Of course, meat is actually some of the body structures of other animals.

Answer the Key Question to the best of your knowledge. Be prepared to share your ideas with the class.

Key Question: Are the proteins that animals eat exactly the same as the proteins that make up their own bodies?

I'm not sure because animals don't usually eat their own kind, so maybe since they're eating different animals they're eating different proteins.

Teacher Talk and Actions
What do we know and what are we trying to find out? (3 min)
Whole class: Read text under "What do we know and what are we trying to find out?" aloud to the class.
Key Question (5 min)
Individuals: Students respond to the Key Question: **Are the proteins that animals eat exactly the same as the proteins that make up their own bodies?** Whole class: Lead discussion. Invite student volunteers to share ideas about the Key Question. The goal is primarily eliciting and probing student ideas here. When students share their answers, do not indicate if their answers are right or wrong. Make it clear that students are expressing ideas, but those ideas may change based on what they will see in the lessons. Challenge any ideas that are inconsistent with learning from previous lessons.

Activity 1: Comparing Proteins in an Animal's Food and Proteins That Make Up the Animal's Own Body

The food that an animal eats often looks quite different from the animal itself. In some cases, the differences are easy to observe. For example, a nursing puppy drinks only milk but needs to make new body structures as it grows. Some species of African snakes eat only eggs, yet they need to make body structures that look quite different from the egg. How is this possible? You know that animal body structures are mainly proteins, so it makes sense for us to examine the proteins that make up animal body structures and compare them with the proteins that make up an animal's food.

Procedures and Questions

1. Observe the photo of the nursing puppy and compare what the puppy eats with the body structures it needs to build.

2. Observe the video of the African snake eating an egg and the photo of the snake shedding its skin. Compare what the snake eats with the body structures it needs to build.

3. Examine Table 4.4 and use the information to answer the questions below.

Table 4.4. Several Proteins, Where They Are Found in the Body, and Some of Their Properties

Protein Name	Body Structures	Solubility in Water	Does It Form Fibers?
Casein	Milk	High	No
Collagen	Tendons, skin	Low	Yes
Hemoglobin	Blood	High	No
Keratin	Skin, scales, hair	Very low	Yes
Myosin	Muscle	Low	No
Actin	Muscle	Low	Yes
Ovalbumin	Egg white	High	No

a. Are milk and muscle made up of the same proteins? Use evidence to support your answer.

No, because they have different properties; for example, casein is highly soluble in water, while myosin is not.

b. Are egg white and skin/scales made up of the same proteins? Use evidence to support your answer.

No, because they have different properties; for example, keratin forms fibers, while ovalbumin does not.

SE p. 149

Teacher Talk and Actions	
Activity 1	**(15 min)**

Show photos of nursing puppies to the students, and allow them time to think about Question 1.

Play the video of the egg-eating snake and show the slides of snakes shedding their skin and allow them time to think about Question 2.

Small groups: Have students work with their group to complete Step 3. Students should be generalizing ideas from Chapter 1 (different substances have different properties, new substances form in chemical reactions) to the context of proteins in animals.

Have students review their notes from Chapter 1 if they are having trouble remembering what solubility is.

Lead a brief class discussion by asking some groups to share their answers.

c. Can you think of any animals that consume mainly hemoglobin but need to make actin and myosin? List them.

Vampire bats, female ticks, spiders, etc.

d. Can you think of any animals that don't consume collagen but need to make it? List them.

Any animal that has skin, such as cows, deer, panda bears (in other words, animals that don't eat meat)

e. Think about the fact that the human body is 50–60% water. Would proteins with high or low solubility in water be better for building human body structures? Explain.

Proteins that have low solubility in water would be the best for building structures because proteins with high water solubility probably would just dissolve in all the water in the body.

f. Are all the proteins listed in Table 4.4 the same substance? Why or why not?

No, they are all different substances because they have different properties.

Teacher Talk and Actions
Activity 1 (cont)
Whole class: Encourage students to give specific examples for Step 3f.

Activity 2: Understanding Proteins

In Activity 1, we saw that different proteins have different properties. We know that different substances have different properties because they are made up of different molecules. But if casein, hemoglobin, myosin, and collagen are all different substances, why do we call them all proteins? In this activity, you will examine the molecular structure of a piece of protein and use what you observe to explain why different proteins have different properties. This activity will also help you think about whether the proteins that are in an animal's food are the same as or different from the proteins that make up its body.

Procedures and Questions

In earlier chapters, you observed several different polymers and the monomers that make them up. You saw that both nylon and Kevlar polymers are made up of two different monomers. In the plant growth chapter, you saw that both starch and cellulose are made up of a single type of monomer—glucose. However, even though both starch and cellulose are made up of the same monomer, they have different properties because the glucose monomers are linked differently. In other words, simply changing the arrangement of the same monomers resulted in starch and cellulose having very different properties.

What about proteins? It turns out that proteins are also polymers. But there are thousands of different proteins. How is this possible? To figure it out, we need to examine the monomers—called amino acids—that make up protein polymers and how they are arranged in different proteins.

1. In Figures 4.1 and 4.2 (p. 152), color the O atoms red and the N atoms blue. Examine the two figures of molecules and answer the questions that follow.

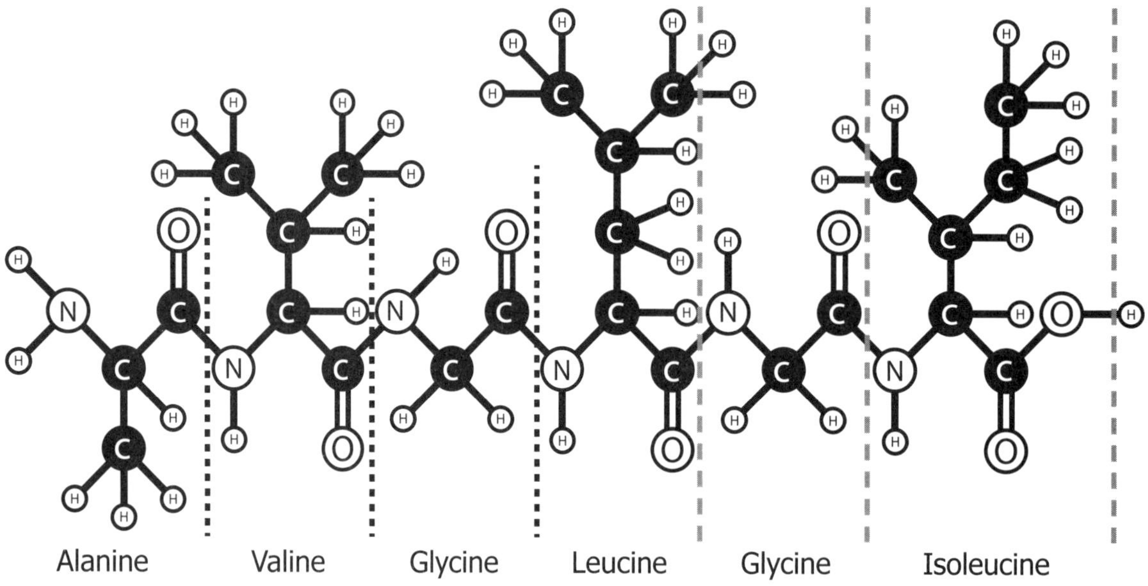

Figure 4.1. Model of the Structure of a Small Piece of a Protein Molecule

Teacher Talk and Actions

Activity 2 **(15 min)**

Figures 4.1 and 4.2 are flattened models that are designed to maximize the visibility of the NH_2 and COOH groups common to all amino acids and the C-N link in the piece of protein.

Students can ignore the double bonds, though they can be helpful for quickly spotting the COOH groups in the amino acids and the CO in the piece of protein.

Take time to make sure all students see the NH_2 and COOH groups in the flattened models of the amino acids and the C-N link in the piece of protein. This may help them find those groups in the 3-D models of proteins they will be working with in Lesson 4.3.

Some students might assume that amino acids are dangerous chemicals because "acid" is in their name. If this comes up, tell students that not all acids are dangerous—acetic acid is vinegar, which people eat and which students have worked with in previous chapters.

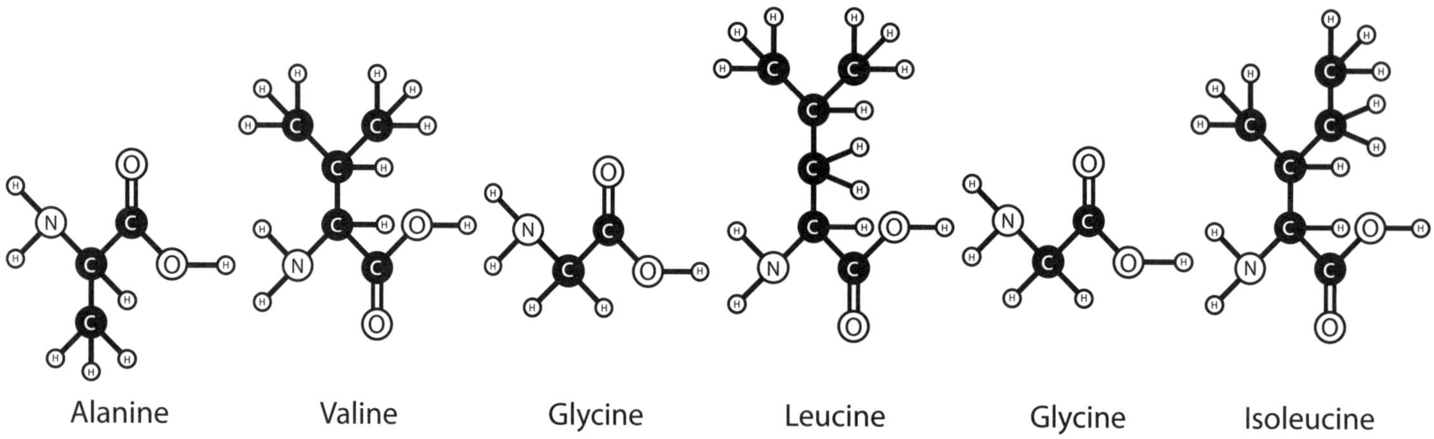

Figure 4.2. Flat Illustration of the Structure of the Amino Acids That the Piece of Protein in Figure 4.1 Would Be Made From

a. What do all the amino acids in Figure 4.2 have in common? Hint: Look at the left and right ends of each molecule.

All the amino acids have a nitrogen atom with two hydrogen atoms (NH_2) on one end and a carbon with two oxygen atoms and a hydrogen atom (COOH or CO_2H) on the other end.

b. What makes some of the amino acids in Figure 4.2 different from each other?

The amino acids differ in the other atoms. Most of those atoms are carbon and hydrogen, but the amounts and arrangements of them differ.

c. Now look at the piece of protein shown in Figure 4.1. The first three amino acid monomers have dotted lines between them and are labeled with their names (*alanine*, *valine*, and *glycine*). Can you find the others? Draw dotted lines between the last three amino acid monomers in Figure 4.1. Hint: Look for parts of the structure that are similar on the left and right ends of each monomer.

d. Do any of the amino acids in Figure 4.1 appear more than once in this piece of protein? If so, which one(s)? Hint: Do you see any from the first three somewhere in the last three?

Yes, glycine

e. Write the amino acid sequence of the piece of protein, using either the names of the amino acids or the first three letters of the names.

Alanine, Valine, Glycine, Leucine, Glycine, Isoleucine

Teacher Talk and Actions
Activity 2 (cont)
Questions 1a–e guide students through identifying key parts of the molecular structure of protein polymers and amino acid monomers. Pay attention to whether students are using "atom" and "molecule" and "monomer" and "polymer" correctly as they discuss the structure. <u>Science Note to Teacher:</u> Proteins are made up of hundreds of amino acid monomers, but there are only 20 types of amino acids, so most will be repeated many times within the sequence that makes up the protein.

f. Now, write at least two other sequences that could be made up of the same amino acids.

Glycine, Valine, Leucine, Glycine, Isoleucine, Alanine

Leucine, Alanine, Isoleucine, Glycine, Valine, Glycine

g. We just looked at several amino acids. Most proteins are made up of 20 types of amino acid monomers that can be arranged to form thousands of different proteins. The protein ovalbumin in egg white is made up of 385 amino acid monomers, and actin, the protein found in muscle tissue, is made up of 374 amino acid monomers. Both proteins contain one or more of all 20 types of amino acid monomers. The sequence of the first 10 amino acids in each protein is shown below. (Only the first three letters of each amino acid name are shown.)

Ovalbumin (from hen egg white):
Met-Gly-Ser-Ile-Gly-Ala-Ala-Ser-Met-Glu ...

Actin (from rabbit muscle):
Asp-Glu-Thr-Glu-Asp-Thr-Ala-Leu-Val-Cys ...

Describe similarities and differences in the partial amino acid sequences of ovalbumin and actin.

The two proteins have some of the same amino acids, such as Ala and Glu. They have many amino acids that are different. The sequence of the amino acids is different in the two pieces we can see.

h. We know that ovalbumin and actin have different properties. Explain, using atoms and molecules, why these two proteins have different properties.

The arrangement of atoms is different in actin and ovalbumin. Different arrangements of atoms means that the two molecules are different substances. We can tell that they are different substances because they have different properties.

Teacher Talk and Actions

Activity 2 (cont)

Science Note to Teacher:

To make most proteins, the body needs to have on hand all the amino acids it needs. If one of the amino acids is in short supply, the body can't make the complete protein. So humans need to eat "complete" proteins or combinations of proteins with all the needed amino acids. This detail was omitted from the Student Edition to keep the storyline as simple as possible.

Read the text of Step 1g aloud to the class.

You might want to write the two sequences up on the board and have students come up and point out any differences/similarities they see.

Have groups write a response and have a brief class discussion.

Students should invoke the earlier science idea that different substances are made of different molecules with different arrangements of atoms.

Science Ideas

Lesson 4.2 was intended to help you understand some important ideas about proteins. Read the idea below. Remember that because science ideas are consistent with evidence from a wide range of examples, you are justified in applying them to new observations and data about similar examples. Look back through Lessons 4.1 and 4.2. Give at least one piece of evidence from your work so far that illustrates each part of this idea.

Science Idea #15: Proteins are the main polymers making up animal body structures. Protein polymers are molecules made of amino acid monomers. Different proteins have different properties because they are made of different types, numbers, and sequences of amino acid monomers.

Evidence:

Table 4.2 in Lesson 4.1 showed us many different kinds of animals, including brine shrimp, chickens, humans, and worms, and all of their bodies were made up of more protein than carbohydrate or fat.

Table 4.4 in Lesson 4.2 showed us that different kinds of proteins have different properties. For example, hemoglobin is soluble in water but keratin is not, so they must be different substances. Activity 2, Question 1g showed us that the order of amino acid monomers is different in different proteins, which also shows us they are different substances.

Teacher Talk and Actions
Science Ideas **(5 min)**
Read the science idea aloud to the class. Give students time to write down at least one piece of evidence that supports each sentence of the science idea. Remind students that they are adding to the list of science ideas students will use in constructing scientific explanations and arguments. Add the science idea to the running classroom list of ideas. As you do, point out that Science Idea #15 is really just an example of science ideas already encountered. Ask students to look at the list of science ideas and identify the science ideas that Science Idea #15 is an instance of. Students should be able to identify Science Ideas #1 and #4.

Pulling It Together

Answer individually and be prepared for a class discussion.

1. Now that you've investigated a variety of proteins, how would you answer the Key Question, **Are the proteins that animals eat exactly the same as the proteins that make up their own bodies?** Answer the question and explain your answer using science ideas, evidence, and models.

The proteins that animals eat are often different from the proteins that make up their own bodies. If substances/proteins have different properties then they are different substances/proteins (Science Idea #4/Science Idea #15). My evidence is from the egg-eating snake, which eats only ovalbumin but makes keratin to replace its shedding skin. Table 4.4 shows that ovalbumin is very soluble in water and does not form fibers, while keratin is not soluble in water and does form fibers. Since ovalbumin and keratin have different properties, they are different substances. Models of amino acids showed different arrangements of C, H, O, and N atoms; so different sequences of amino acids should also have different arrangements of atoms. I'm not sure how animals can make different proteins from the ones they eat, but they must be able to do so. Maybe animals can change the sequence of amino acids, which would make the proteins different.

Teacher Talk and Actions

Pulling It Together (10 min)

Individuals: Students answer Pulling It Together questions on their own, in detail.

Whole class: Lead a class discussion.

Invite students to share ideas about Pulling It Together Question 1, which returns to the lesson Key Question. Direct students to look back at their initial ideas about the Key Question and then consider how their ideas may have changed.

Science Note to Teacher:

Some species of animals are occasionally cannibalistic and eat members of their own species. In this case, the animal doing the eating is actually eating many of the exact same proteins that make up its body. However, the proteins are still broken down into amino acids and transported to cells where they are used to build more proteins. Proteins aren't just "tacked on" "as is" to animal body structures. The Student Edition avoids discussion of this as it requires somewhat advanced knowledge of how animal body systems connect and relate and only serves to explain an exception to the rule.

2. Look back at the questions at the end of Activity 1. How do you think the egg-eating snake from Activity 1 might be able to get actin for building its muscles and keratin for building its scales from ovalbumin?

Maybe the snake can use the amino acids from ovalbumin and build proteins that have different amino acid sequences.

3. Most meats that humans eat are the muscles of other animals (cows, chickens, pigs, fish, and turkeys). When people prepare meat for meals, they typically use only the muscle. These muscles are made up mostly of the proteins *actin* and *myosin*. But when people grow, their bodies need to build more than just muscle. They need to build tendons (mostly collagen), skin (mostly collagen and keratin), and hair and nails (mostly keratin). What could explain how these different proteins could be made from actin and myosin?

Humans must be able to make proteins like collagen and keratin that have different amino acid sequences from the proteins they eat.

Maybe a role of the digestive system is to produce amino acid monomers from protein polymers.

Teacher Talk and Actions

Pulling It Together (cont)

Pulling It Together Question 2 expects students to apply the idea that if different proteins are made of different amino acids (Science Idea #15) then maybe the snake can use amino acids from one protein (e.g., ovalbumin) to make another (e.g., keratin).

Pulling It Together Question 3 expects students to recognize that both are polymers made up of monomers but the number of types of monomers is different. Students should be thinking about digestion but aren't expected to know that the proteins are digested to monomers or that the monomers are used to build body proteins.

Closure and Link (2 min)

Whole class: Summarize the lesson in a few statements.

We saw that proteins are polymers, made from amino acid monomers. We saw that there are many types of proteins, which have different properties. We also saw that animals often eat different proteins from the proteins that make up their own body structures, which they need to make as they grow and repair.

In the next lesson, we'll see what happens when animals eat proteins in food. You will have the chance to model what happens to help you explain how animal bodies build the proteins they need to grow and repair.

Lesson Guide
Chapter 4, Lesson 4.3
Explaining Animal Growth With Atoms and Molecules

Focus: This lesson focuses on how protein molecules in food become part of animal body structures. It uses experimental evidence and models to show how protein polymers are broken down into amino acid monomers in the digestive system and then new protein polymers are built from those amino acid monomers. It confronts the misconception that food either leaves the body as waste or is burned for energy by providing evidence that some atoms from food are incorporated into animal bodies to support growth and repair.

Key Question: How do animals use proteins from food to repair and build their body structures?

Target Ideas Addressed

- Growth and repair of animal body structures involve chemical reactions during which proteins from food are used to make other proteins that become part of animals' body structures.
- The process by which proteins from food become part of animals' body structures involves chemical reactions in which the proteins from food are broken down into amino acid monomers, and these monomers are used to build different protein polymers that make up body structures.
- When animals grow or repair, they increase in mass. Atoms are conserved when animals grow: The increase in measured mass comes from the incorporation of atoms from molecules that were originally outside of the animals' bodies.

Science Practices Addressed

- Analyze and interpret data to provide evidence for phenomena.
- Use multiple models to describe phenomena.
- Apply scientific ideas, principles, or evidence to construct an explanation for real-world phenomena.
- Read scientific texts adapted for classroom use to determine the central ideas or obtain scientific or technical information to describe patterns in or evidence about the natural world.

Materials

Activity 1: Projector system; Slide show: *Herring Fish and Aquaculture*, Video—*Brine Shrimp*

Activity 2: Ball-and-stick model kits, Stickers for models, Markers (dry or wet erase) for writing on clear mat covers, Tray for working with models on desks, Copymasters: *Chemical Reaction Mat:* ◆ Reaction #1: Breaking Down a Piece of a Protein Into Amino Acids; *Chemical Reaction Mat:* ◆ Reaction #2: Building a Piece of a Protein From Amino Acids; *Chemical Reaction Mat:* ■ Reaction #1: Breaking Down a Piece of a Protein into Amino Acids; *Chemical Reaction Mat:* ■ Reaction #2: Building a Piece of a Protein from Amino Acids

Advance Preparation

This is a very long lesson, so depending on your class period length, look at the lesson phase time estimates (on the teacher facilitation pages) and plan how you will split it up.

Activity 1: Set up the projector and locate the video and slide show on the *THSB* website.
Activity 2: Before class—For each group of four students, use the ball-and-stick kits to build four

water molecules and one piece of protein shown on the *Chemical Reaction Mat* ■ Reaction #1: Breaking Down a Piece of a Protein into Amino Acids.

- Stick a sticker on one of the carbon atoms on each of the protein pieces you make.
- The piece of protein students start with will have the same five amino acids as the piece of protein they end with, but in a different order, to show it is a part of a different protein. The next class will start with the last class's ending protein. This means that Reaction #1 for the next class will be a different mat. For each section of students you have, you will alternate which set of mats, designated (■ or ◆) students will be using. The activity is designed this way to avoid having to build new models between classes. Although you will have to check that students have built their ending protein correctly, you should not have to build models from scratch between class periods.
- Practice the modeling activity that students will be doing, and then try starting with the ending materials and the alternate set of mats to do what the next class of students will do. Practicing with the models is key because it can be hard to see mistakes and help students fix their models unless you are highly familiar with the activity.

Phenomena, Data, or Models	Intended Observations	Purpose	Rationale or Notes
Activity 1 Data from a radioactive labeling experiment that shows how much protein from food ended up in the gut, in the rest of the body, and as waste when herring fish ate brine shrimp with labeled proteins	When herring are fed brine shrimp with labeled proteins, atoms from the proteins first enter the fish's digestive system, and then over time these atoms leave the digestive system and end up as either waste or as part of the fish's body.	Students are shown evidence that some atoms from protein molecules become part of the animal's body, and therefore contribute to growth and repair. This evidence is contrary to the common student misconception that all food becomes either waste or is "used up" as energy for growth.	In Table 1, "Waste products" is a sum of catabolized protein (which ends up as carbon dioxide, 39%) and feces/urine (34%). Because this unit does not address energy or cellular respiration, carbon dioxide is lumped into this category as a waste product.
Activity 2 Ball-and-stick modeling activity (with a labeled carbon atom) that represents what happens to proteins in food when an animal uses food it eats to build/repair its body	When animals eat food containing proteins, those protein polymers are broken down to amino acid monomers in the digestive system by chemical reactions that require water molecules. Amino acid monomers are then used to make protein polymers wherever in the body new proteins are needed. The new protein polymers are usually different proteins from the proteins in the food, and therefore have different amino acid sequences.	Students use models to visualize reactions happening inside an animal's body to account for atoms from food becoming part of its body.	The lesson does not explain the role of digestive enzymes such as pepsin and trypsin or of stomach acid in lowering pH for the enzymes to work well. For students just learning the basics of chemical reactions, the catalytic role of enzymes in speeding up chemical reactions would overly complicate the story.

Lesson 4.3—Explaining Animal Growth With Atoms and Molecules

What do we know and what are we trying to find out?

In the last lesson, we saw that the proteins an animal eats are rarely exactly the same proteins it needs to build its body structures. So what happens to food when an animal eats? When a puppy nurses, how does it get the proteins to build its body structures if it mainly takes in casein from its mother's milk? When the egg-eating snake ingests the protein ovalbumin, how does it get the proteins actin and myosin for its muscles and keratin for its skin?

In this lesson, you will have the chance to use experimental data and models to figure out what happens to the proteins in food once an animal has eaten them, and how they help an animal grow and repair its body structures.

Answer the Key Question to the best of your knowledge. Be prepared to share your ideas with the class.

Key Question: How do animals use proteins from food to build and repair their body structures?

Proteins from food go to the stomach (or small intestine/digestive system) and get digested, or broken down into tiny pieces. Then they might end up as waste, or be used for energy, or end up as part of the body. (Note: At this point, not all students are expected to know that atoms from food must end up as part of the body if animals are gaining weight.)

Teacher Talk and Actions
What do we know and what are we trying to find out? (3 min)
Whole class: Read text under "What do we know and what are we trying to find out?" aloud to the class.
Key Question (5 min)
Individuals: Students respond to the Key Question: **How do animals use proteins from food to build and repair their body structures?** Whole class: Lead discussion. Invite student volunteers to share ideas about the Key Question. The goal is primarily eliciting and probing student ideas here. When students share their answers, do not indicate if their answers are right or wrong. Make it clear that students are expressing ideas, but those ideas may change based on what they will see in the lessons. Challenge any ideas that are inconsistent with learning from previous lessons.

Activity 1: Observing What Happens to Food Proteins as an Animal Grows

In Chapter 3, you saw how scientists used labeled atoms to determine the reactants and products of chemical reactions that take place in plant body structures. The labeling technique was useful because scientists could prepare reactants with some labeled atoms and because the labeled atoms rearranged just like normal atoms do. You used data from these experiments to show that the carbon and oxygen atoms of the glucose molecules plants make come from carbon dioxide in the air. Do you think this technique can help us figure out how the egg-eating snake gets keratin, actin, myosin, and lots of other proteins when it eats mostly ovalbumin? Can it help us figure out how the puppy gets actin, myosin, hemoglobin, and lots of other proteins when it eats mostly casein in its mother's milk? In this activity, you will examine an experiment to help you answer these questions.

Procedures and Questions

1. Read about the experiment that scientists conducted. Examine the data they collected, and then answer the questions that follow with your team. Think about what the experiment can tell you about what happens to food when an animal eats. Observe the photos that your teacher will show you to help you understand the experiment.

Brine Shrimp Are a Suitable Food Source for Farming Herring

Introduction

Aquaculture is the farming of fish and other organisms that live in water, just as agriculture refers to the farming of land organisms. Scientists study fish in aquaculture to determine the best ways to grow the most fish. Part of this involves studying the food that is fed to the fish and how much fish grow when they are fed different foods. Scientists had previously found that catfish could be farmed by feeding them brine shrimp, tiny crustaceans that are relatives of shrimp people eat. In our study, we wanted to know if herring could also be farmed by feeding them brine shrimp.

Methods

We determined how to incorporate labeled carbon atoms into proteins in brine shrimp. We then fed these brine shrimp to young herring. We first measured how long the labeled carbon atoms from the brine shrimp proteins remained in the digestive system of each fish. Once we knew when those carbon atoms were out of the digestive system, we could see where they went. Data are shown in Figure 1.

Teacher Talk and Actions	
Activity 1	**(35 min)**
Whole class: Read the first paragraph under Step 1 aloud to the class. Project the images of herring and brine shrimp to help students visualize the experiment.	

Results

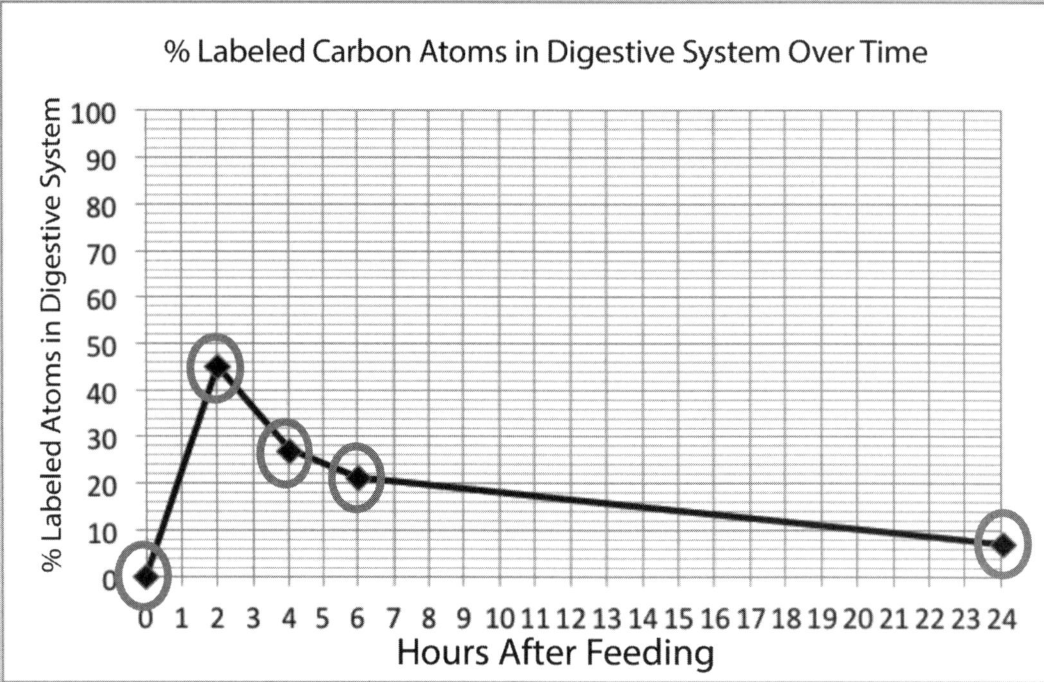

Figure 1. Percentage of labeled carbon atoms from food in a fish's digestive system over time (Time 0 was before feeding)

a. At which time points did the scientists take measurements? Circle them on Figure 1.

b. After feeding,

i. when was the percentage of labeled atoms in the fish's digestive system the highest?

2 hours after feeding

ii. when was it the lowest?

24 hours after feeding

c. What does Figure 1 tell you about the amount of food in the digestive system *over time*?

The food in the digestive system initially increased (right after feeding) and then slowly decreased. After 24 hours, very little food (7%) was left in the fish's digestive system.

Activity 1 (cont)

Small groups: Have students read Figure 1 and answer Questions 1a–c. Make sure students understand that food has entered and then left the digestive system of the fish.

Make sure students understand that if scientists find radioactive atoms in a different substance than the substance that originally had those atoms, then atoms must have rearranged and, therefore, a chemical reaction must have occurred.

Twenty-four (24) hours after feeding, we measured the amount of labeled carbon atoms in three places: (1) in the water, (2) in the digestive system of each fish, and (3) in the rest of its body (as shown in Figure 2).

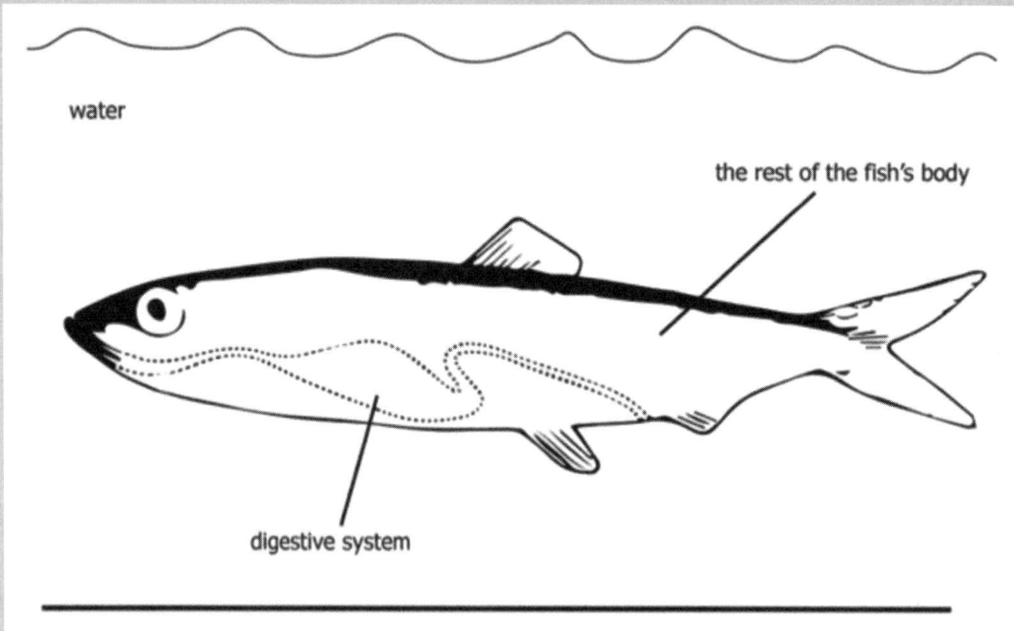

Figure 2. Representation of a herring in an aquaculture tank

Table 1. Percentage of labeled carbon atoms found in three locations 24 hours after feeding the fish with food containing labeled carbon atoms

Location	% of Labeled Carbon Atoms Found 24 Hours After Feeding
Digestive system	7
Rest of body (excluding digestive system)	20
Waste products (in the water, outside the fish's body)	73

Morais, S., L. E. C. Conceicao, M. T. Dinis, and I. Ronnestad. 2003. A method for radiolabeling Artemia with applications in studies of food intake, digestibility, protein and amino acid metabolism in larval fish. *Aquaculture* 231 (1–4): 469–487.

d. Remember that we are trying to find out what happens to the proteins in the food that an animal eats. What does Table 1 tell you about where the labeled atoms ended up once they left the fish's digestive system?

A lot of the labeled atoms ended up as part of waste products, some of them ended up as part of the body of the fish, and a few of them were still in the digestive system of the fish.

Teacher Talk and Actions

Activity 1 (cont)

Table 1 shows the different fates of the food once it has left the digestive system.

Science Note to Teacher:

Scientists dissected the fish to separate the guts (digestive system) from the rest of the body when measuring the amount of radioactive atoms in each "compartment."

e. Did all of the labeled atoms end up as waste products? Give evidence for your answer.

No—some of them (20%) ended up as part of the body. My evidence is that Table 1 says that 20% were in the rest of the fish's body.

f. When a fish eats brine shrimp, is it more like what happens when iron rusts in an open container or when baking soda reacts with vinegar in an open container? Explain your answer using science ideas from Chapter 2.

The fish eating is more like iron rusting. A fish's body is like an open container, because we are adding food to the body. Science Idea #10 says that the measured mass of reactants and products is not always the same as the total mass. The measured mass changes if reactants or products (often gases) enter or leave the system. This is because atoms that make up reactants or products enter or leave the system. When food atoms enter the system, the total mass increases.

g. What do you think happened to the mass of the fish 24 hours after feeding—that is, 24 hours after time point 0? Why do you think so?

I think the mass of the fish increased because it ate food. Food is matter, so it is made of atoms. The fish is eating the food, so it is adding atoms to itself. The atoms have mass, so the mass of the fish increases.

h. What will happen to the fish's mass if it continues to eat brine shrimp for many days?

The mass of the fish will continue to increase if the fish continues eating brine shrimp for many days. The food it eats has atoms, and therefore, mass. The fish will continue to add mass to itself.

Teacher Talk and Actions
Activity 1 (cont)
Small groups: Have students answer the questions. Remind students of the Key Question: How do animals use proteins from food to build and repair their body structures? How can these data help answer that question? <u>Science Note to Teacher:</u> In Table 1, "waste products" looks like a high number because it is the combination of waste (feces/urine; 39%) and carbon dioxide (product of cellular respiration; 34%). Both are released into the water.

Activity 2: Modeling the Breakdown and Building of Proteins in Herring

Materials

For each team of students

Prebuilt ball-and-stick models of four water molecules and a piece of a protein with a labeled carbon atom

Chemical Reaction Mats: Reaction #1: Breaking Down a Piece of a Protein Into Amino Acids, Reaction #2: Building a Piece of a Protein From Amino Acids

Markers for writing on clear mat covers

Tray for working with models on desks

In Activity 1, you saw some data on where the carbon atoms in food go when it is eaten by an animal—in this case, herring. The data showed that some of the labeled carbon atoms ended up in the fish's body structures. You know that herring have a lot of muscle that they need to build in order to grow. Can the herring just take the protein from the brine shrimp and add it to their muscles unchanged? Or do chemical reactions play a role?

In Activity 2, we will use models to help answer that question. We will start with a piece of protein with a labeled carbon atom on a single amino acid and note where that amino acid is in the protein. Then we will model two chemical reactions—protein digestion and protein synthesis—that are involved when animals take in and use food to build and repair their bodies. Our models will help us see whether the labeled amino acid in the reactant molecules ends up in the same place in the amino acid sequence of the product molecules.

Procedures and Questions

Complete the steps in the procedure and respond to the questions. Steps 1–8 focus on the process by which proteins in food are digested in an animal's body. Steps 9–16 focus on the process by which new proteins are made (synthesized).

1. As a team, examine your model of a piece of food protein. Find the carbon atom that has a sticker on it. The sticker represents the label on the carbon atom. Write down the amino acid sequence of the piece of protein and highlight the amino acid that has the label. Where in the amino acid sequence is the labeled amino acid?

   Answer will vary.

Teacher Talk and Actions

Activity 2 **(40 min)**

Whole class:

Read the text under Activity 2 aloud to the class.

To each group of 3–4 students: Pass out the *Chemical Reaction Mat* for Reaction #1: Breaking Down a Piece of a Protein into Amino Acids (■ or ◆ depending on the class period), a premade model of a piece of protein, and four water molecules. Make sure each group has a marker for writing on the mats.

Small groups:

Have groups work through Steps 1–4. At Step 4, have them wait for you to check their "plans" (marked on the mat) before they start to manipulate the models. (A correctly marked mat is shown.)

If students struggle to decide where the breaks will occur, have them look back at Figures 4.1 and 4.2 (Lesson 4.2, pp. 151–152) to remember the basic structure of proteins and amino acids. Direct them in particular to the dotted dividing lines in Figure 4.1.

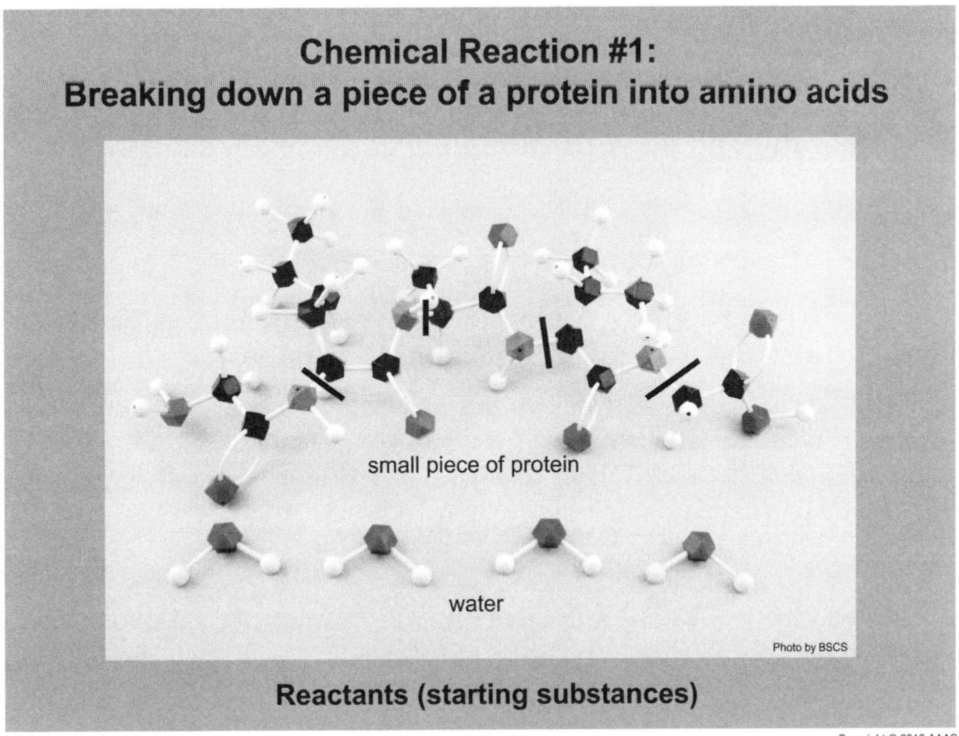

2. Look at the yellow side of the *Chemical Reaction Mat* for Reaction #1, which shows what the products of the chemical reaction will be. How many amino acids will this piece of food protein break down into?

   Five amino acids

3. How many connections between atoms in the food protein will need to break in order to make the amino acids?

   Four connections

4. Using the markers provided, draw lines on the green side of the mat through the connections your team plans to break in the piece of food protein. (Ask your teacher for help if you are having trouble deciding. All teams should check with their teacher before moving on to Step 5.)

5. Assign each team member a connection to break. Put your initials by the line through the piece of food protein that you will be breaking.

6. Flip your *Chemical Reaction Mat* to the yellow side (*Products*), remove the yellow sheet from the plastic sleeve, and place it next to the green side (*Reactants*).

7. Using only your piece of food protein and water molecules, build models of the products as shown on the yellow side of the mat. After you've finished building the products, **do not take apart your models.** Put the yellow sheet back in the sleeve.

8. With your team, answer the following questions.

   a. You started with a small piece of a protein polymer and four water molecules. What happened to the water molecules during the chemical reaction?

   The atoms from the water molecules made new connections and became part of the amino acid molecules. (A few students might indicate that the H and O atoms are now part of the NH_2 and COOH groups. This should become evident to others in the next activity).

   b. Now that you've made the products, are any atoms connected to different atoms than they were before? If so, give an example.

   Yes. The O atoms that were connected to two H atoms are now connected to carbon atoms and one of each of the H atoms is now connected to N.

Teacher Talk and Actions
Activity 2 (cont)
Small groups: Have groups work through Steps 5–7 and the questions under Step 8. Tell students not to take apart their models when they're done with Step 8. Remind students to refer back to the data in Activity 1 to recall the real-life phenomenon they are modeling. When you see that a group is done with the questions in Step 8, give it the *Chemical Reaction Mat* for Reaction #2: Building a Piece of a Protein From Amino Acids (■ or ◆ depending on the class period). Tell students that the products of their first reaction are the reactants of the second reaction.

c. Is the label on the same carbon atom as at the start of the chemical reaction? Is the labeled carbon atom part of the same molecule as it was at the start of the chemical reaction?

The label is on the same carbon atom. It's on an amino acid now, when before it was on a piece of protein. Amino acids have a different number and arrangement of atoms, so they have different properties from each other, too.

d. Do you think a protein and an amino acid will have the same properties? Why or why not?

No. Science Idea #4 tells us that different number, type, and arrangement of atoms means different substances with different properties. In our piece of protien, there are 18 carbons, where in the biggest amino acid there are only 6. There are similar differences for oxygen, hydrogen, and nitrogen atoms. Because the number of atoms is different, they must be different substances.

e. Did you gain any atoms or lose any atoms during this chemical reaction?

No—we have the same number we started with.

f. Write a word equation for the reaction you just modeled.

Protein molecule + water molecules → amino acid molecules

g. Where in the fish's body do you think Reaction #1 happens?

Somewhere in the digestive system

9. Your teacher will give you a new *Chemical Reaction Mat* for Reaction #2.

10. With your team, examine Reaction #2. The products you made in Reaction #1 should match the pictures on the green side of the mat for Reaction #2.

11. Remove the yellow side of the mat for Reaction #2, which shows the products of the chemical reaction, and place it next to the green side. Using the green side to identify the amino acids, write down the amino acid sequence of the protein product (yellow side).

Will either be:

Glycine, Leucine, Alanine, Valine, Glycine OR

Alanine, Glycine, Valine, Glycine, Leucine

Teacher Talk and Actions

Activity 2 (cont)

Small groups:

Have groups work through Steps 10–12. Before giving them the OK to move on, check their "plans" to be sure that in Step 12 they have shown connections between C and N atoms of adjacent amino acids as shown in the example below. Note that since there are two glycine monomers, there are two ways to connect the five amino acids to make the protein. You might want to find the other way in case your students take that path.

Be sure students do not disassemble their models of the protein product of Reaction #2. You will need these for the next class.

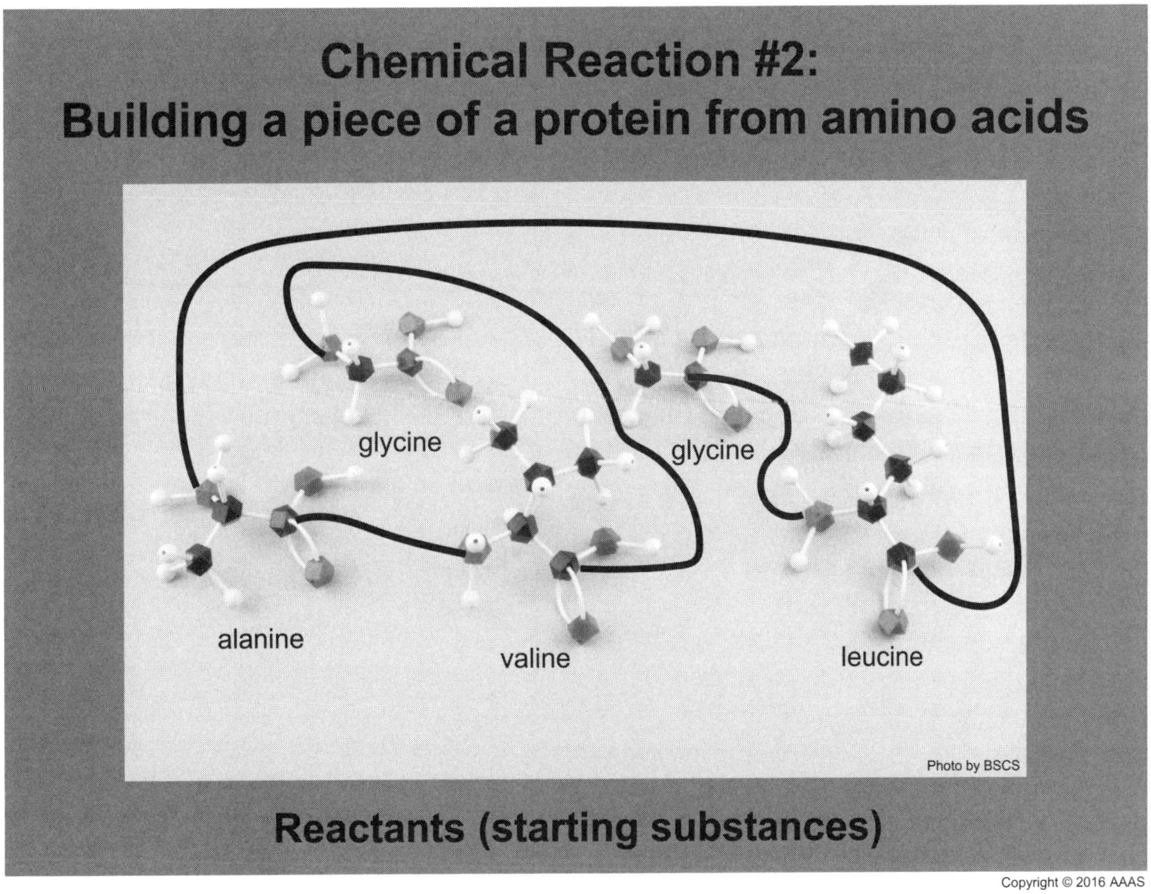

12. Arrange your amino acid models according to the sequence you wrote down, and have your teacher check the arrangement.

13. On the green side of the mat for Reaction #2, circle the atoms that will separate and form new connections.

14. Assign each team member a connection to make. Put your initials by atoms that you will separate and use to form new connections.

15. Build the products for Reaction #2 as shown on the yellow side of the mat. Place the yellow side of the mat back in the sleeve. **Do not take apart your models of the polymer—leave them for the next class.**

16. With your team, answer the following questions.

    a. Write down the amino acid sequence of the piece of protein you just built, and highlight the amino acid that has the label. Where in the sequence is the labeled amino acid?

Answer will vary depending on location of the sticker.

    b. Compare the amino acid sequence of the model of the protein you just produced with the amino acid sequence of the model of the protein you started with in Step 1. Are the proteins you modeled the same or different? Use science ideas and models to explain your answer.

The proteins are different. They are made of different arrangements of amino acids. Science Idea #15 tells us that different proteins have different types, numbers, or arrangements of amino acids. The models showed us that the amino acids are in a different order.

    c. Where in the fish's bodies would the products from Reaction #2 be found? Support your answer with evidence.

In fish body parts—for example, as part of the muscles—anywhere new proteins are being made for growth or repair. My evidence is that Table 1 in the experiment says that 7% of the atoms stayed in the bodies.

Teacher Talk and Actions
Activity 2 (cont)
As groups finish the activity, remind them **not to take apart their final products.** (You, the teacher, need them for the next class period! Check that students have built their proteins correctly as you wait for everyone to finish.) Lead a class discussion by asking some students to share their answers for the questions under Step 8 and then under Step 16. Watch carefully for correct use of *atom, molecule, monomer, polymer, measured mass,* and *total mass* when students share their answers. If they say *mass,* ask, "Measured or total?" If they say *atoms and molecules,* ask, "Which one do you mean?"

d. Let's consider how the two chemical reactions you modeled relate to the herring data. Where were the atoms at the beginning of the experiment (time 0 on the graph)? Where were the atoms after 24 hours?

At the beginning (time 0) the atoms were part of proteins of the brine shrimp. Before the brine shrimp were fed to the herring fish, the atoms were outside the herring fish's body.

After 24 hours, some of the atoms (20%) were now part of the fish's body. That means there would be more atoms in the fish's body after 24 hours compared with time 0.

e. What does your answer for Step 16d mean for the measured mass? Will it change or stay the same? Why?

The measured mass of the fish will increase after 24 hours, because some of the atoms from the brine shrimp became part of the fish's body.

Teacher Talk and Actions
Activity 2 (cont)
Step 16d: Check to be sure students understand that they are comparing time 0 with 24 hours. Some students may think (incorrectly) that the number of atoms after 24 hours is less. This is likely because they are comparing the number of atoms after 24 hours to the number of atoms AFTER feeding, rather than to the number of atoms BEFORE feeding.

Science Ideas

Lesson 4.3 was intended to help you understand some important ideas about how food relates to growth in animals. Read the ideas below. Notice that Science Ideas #16 and #17 explain observations about animal growth in terms of atoms. We can observe when animals grow, but we can't see atoms rearranging or entering or leaving the system. However, because Science Ideas #16 and #17 state general principles that are consistent with a wide range of observations and data, we can use them to reason about the growth of all animals. You will be expected to use ideas about atoms to explain phenomena involving animal growth.

Look back through Lesson 4.3. Give at least one piece of evidence from your work so far that illustrates each of the science ideas listed below.

Science Idea #16: The process by which proteins from food become part of animals' body structures involves chemical reactions in which the proteins from food are broken down into amino acid monomers, and these monomers are used to build different protein polymers that make up body structures. Atoms are rearranged during both the breakdown and the building of protein polymers.

Evidence:

In Activity 2, we modeled how proteins from food could react with water and break down into their amino acid monomers. We saw that balls representing H and OH atoms from water ended up on the ends of the monomers. The H atoms became part of NH_2 groups and the OH became part of COOH groups. The amino acid monomers reacted in a different order to make different proteins. We saw that when the monomers reacted to make different proteins, the balls representing H came off the NH_2 groups and the balls representing OH came off the COOH groups. So our models showed us how a small number of atoms could rearrange during the formation of large polymers.

Science Idea #17: When animals grow or repair, they increase in mass. Atoms are conserved when animals grow: The increase in measured mass comes from the incorporation of atoms from molecules that were originally outside of the animals' bodies.

Evidence:

In Activity 1, we observed that fish taking in food (brine shrimp) incorporate some labeled atoms from that food into their bodies. Since some atoms that used to be outside the body are now a part of the body, the fish would be bigger (have more mass).

Teacher Talk and Actions
Science Ideas **(10 min)**
Read the science ideas aloud to the class. Give students time to write down at least one example that supports the idea(s). Remind students that the science ideas will be used in constructing scientific explanations. Add the science idea(s) to the running classroom list of ideas. As you do, point out that these are simply instances of the science ideas encountered in Chapters 1 and 2 and ask students to identify those science ideas. Students should notice that Science Idea #16 is simply an instance of Science Ideas #2 #5, and #6 and that Science Idea #17 is an instance of Science Ideas #7 and #10. This emphasizes the importance of chemical reactions and matter conservation in both nonliving and living systems.

Pulling It Together

1. After examining the data and models in this lesson, how would you answer the Key Question, **How do animals use proteins from food to build and repair their body structures?**

   Some atoms from proteins from food become part of the animal's body. Through chemical reactions, they are broken down into their amino acid monomers in the digestive system, and the amino acids are then used to build new proteins wherever the body needs them for growth and repair.

2. We used models to reason about how a fish makes its body protein from eating brine shrimp. Now explain, using models, how egg-eating snakes can make proteins like myosin and keratin when they mainly eat ovalbumin.

   The snake eats the ovalbumin, which is a protein, like our piece of protein model. Inside the snake's body, the food is broken down to amino acid monomers, like how we separated the amino acid models. The snake rearranges the amino acid monomers, and makes new connections between them to make new proteins. The new protein could be myosin or keratin.

3. Imagine your friend has just finished eating a turkey sandwich. You can represent the atoms making up proteins in the turkey meat with LEGO bricks. How would the total number of bricks that were in the sandwich before your friend ate it compare with the total number of bricks inside your friend's body immediately after she ate it? Would all the bricks stay inside her body? Would any of the bricks disappear? Support your answer with evidence.

   The total number of LEGO bricks would stay the same, because atoms (represented by LEGO bricks) are not created or destroyed in chemical reactions. None would disappear, but some might leave the body as waste. Some would stay in the body and become part of new proteins that would add to my friend's body during chemical reactions involved in growth or repair.

Teacher Talk and Actions
Pulling It Together **(20 min)**
Individuals: Students answer Pulling It Together questions on their own, in detail. Whole class: Lead a class discussion. Invite students to share ideas about Pulling It Together Question 1, which returns to the lesson Key Question. Direct students to look back at their initial ideas about the Key Question and then consider how their ideas have changed. Students should have an improved answer for Pulling It Together Question 1, the Key Question. If students need help answering Pulling It Together Question 2 correctly, after helping them you can assign this question as practice: Explain how the egg-eating snake can replace its shed skin. Alternatively, you can lead a class discussion on how the egg-eating snake replaces its shed skin and then assign Pulling It Together Question 2 for students to complete independently. Pulling It Together Question 3 requires students to be comfortable shifting between different types of models (it has been a while since they've seen LEGO bricks) and between models and the phenomena they represent.

4. When you get a cut on your hand, your body builds scar tissue to seal up the cut. Scar tissue is made up of the protein *collagen*. Explain how this can happen. Be sure your explanation meets the Explanation Quality Criteria.

The matter to build the new scar tissue comes from atoms from food I eat. Different substances have different properties because they are made up of different arrangements of atoms (Science Idea #4). Collagen has different properties from proteins humans eat. For example, collagen is not water soluble, whereas casein (in milk) and ovalbumin (from egg white) is water soluble. Therefore, collagen has a different arrangement of atoms (or different sequence of amino acids) from proteins we eat. Proteins from food become part of animal body structures during chemical reactions such as protein digestion and protein making. Atoms are rearranged during both the breakdown and building of protein polymers (Science Idea #16). Table 4.2 includes humans in the list of animals, so I can apply Science Idea #16 to humans. When we modeled the digestion of brine shrimp protein to amino acids and then used those amino acids to build fish protein, we saw that amino acids were sequenced differently. The modeling showed us that it was possible to make different proteins from proteins in food. So I think humans build collagen for scar tissue by digesting proteins from food to amino acids and using those amino acids to build collagen.

5. Describe how the addition of new protein molecules contributes to the growth of an animal's body structures and body.

New protein molecules are made by digesting food proteins to amino acids and using the amino acids to build body proteins (Science Idea #16). Some atoms from food leave the body as waste, but others become part of the body. When animals incorporate atoms from food into their bodies, they increase in mass (Science Idea #17). My evidence is the data from the fish experiment, which showed that 24 hours after eating radioactively labeled brine shrimp, 20% of the labeled atoms had been incorporated into the herring fish's body structures. As an animal continues to grow, it incorporates more and more atoms from food into its body. Since atoms have mass, more atoms making up the body means the body has more mass.

Teacher Talk and Actions

Pulling It Together (cont)

Pulling It Together Question 4 reminds students to be sure their explanation meets the Explanation Quality Criteria. Point out that they are expected to do this whenever they are asked to construct an explanation, regardless of whether or not they are reminded.

Students may list Science Idea #16 but not justify why it is relevant. Challenge students by asking them, "why do you think it's OK to use Science Idea #16? Does the idea apply to humans? What evidence can you cite?"

Check to be sure students understand that even though the modeling activity doesn't provide evidence that humans turn food proteins into collagen, it does show that it is possible to do so by carrying out chemical reations.

Pulling It Together Question 5 is intended to get students to realize that more atoms (as part of monomers making up protein molecules) making up the body means a bigger, more massive body. Note: Though the molecules aren't broken down to individual atoms, it is most useful to think about the total number of atoms (rather than molecules) when determining if mass has increased or not (as they experienced in Chapter 2).

Closure and Link (2 min)

Whole class:

Summarize the lesson in a few statements:

We've learned that when animals eat, some of the atoms from proteins in food become part of the animal's body. We used models to see how this happens—inside animal bodies, protein polymers react with water molecules and are broken down into amino acids. This happens in the digestive system. The amino acids enter the blood stream where they are carried throughout the body. These amino acids can react in different orders to form new proteins and water. This happens wherever the body needs new proteins for growth or repair of body structures.

Link to next lesson:

In the next lesson, you will examine how scientists use evidence and reasoning to justify their claims about muscle growth and repair. You will have the chance to apply everything you've learned in Chapters 1, 2, and 3 to develop a scientific explanation of growth in animals.

Lesson Guide
Chapter 4, Lesson 4.4
Examining Explanations of Animal Growth and Repair

Focus: This lesson engages students in applying ideas from Chapters 1, 2, and 3 to analyze, construct, and refute arguments about explanations of animal growth.

Key Question: How are the explanations we have been writing similar to the explanations that scientists write when they publish their work?

Target Ideas Addressed

- Students should be able to use all the science ideas from Chapters 1, 2, and 3.

Science Practices Addressed

- Analyze and interpret data to provide evidence for phenomena.
- Apply scientific ideas, principles, or evidence to construct an explanation for real-world phenomena.

Materials

None

Advance Preparation

Review student work in Chapters 1, 2, 3, and 4 and be prepared to help students find relevant evidence and ideas in their notebooks.

Phenomena, Data, or Models	Intended Observations	Purpose	Rationale or Notes
Activity 1 Experimental data on the amount of ^{14}C-myosin produced by treatment group (complete meals + amino acid supplement) and control group (complete meals)	The treatment group produced high levels of radioactive myosin, whereas the control group produced very low levels of radioactive myosin.	Students consider the evidence provided to support the claim that taking an amino acid supplement causes elderly men to increase protein production (though not necessarily a net increase in their muscle mass).	The activity provides an opportunity for students to examine how scientists support claims with evidence and control for variables.
Pulling It Together Students cite evidence to support the claim that chemical reactions happen in animals, not just in the lab.	Data on mass increases in growing animals, animal body composition, protein properties, and radioactive labeling experiments provide evidence that animals make new substances as they grow. Modeling activities show that it is possible to make different proteins by rearranging atoms from proteins in food. Hence, chemical reactions occur in animals as they grow.	Students have the opportunity to synthesize the phenomena and modeling from Chapter 4 lessons to support the claim that animal growth involves chemical reactions.	The question provides another opportunity for students to link evidence to claims.

Lesson 4.4—Examining Explanations of Animal Growth and Repair

What do we know and what are we trying to find out?

In Chapter 4, you have investigated how animals grow. You have applied what you learned about atoms, molecules, chemical reactions, and mass in Chapters 1, 2, and 3 to explain animal growth. This lesson will give you a chance to examine and evaluate other people's explanations of animal growth.

Answer the Key Question to the best of your knowledge. Be prepared to share your ideas with the class.

Key Question: How are the explanations we have been writing similar to the explanations that scientists write when they publish their work?

Scientists' explanations probably need to meet the Explanation Quality Criteria we've been using. For example, scientists need to support their conclusions with evidence. Maybe they don't need to state the science ideas, if their conclusions are consistent with them. However, if their conclusions are not consistent with accepted science ideas, they need to try to explain why.

Teacher Talk and Actions
What do we know and what are we trying to find out? (3 min)
Whole class: Read text under "What do we know and what are we trying to find out?" aloud to the class.
Key Question (5 min)
Individuals: Students respond to the Key Question: **How are the explanations we have been writing similar to the explanations that scientists write when they publish their work?** Whole class: Lead discussion. Invite student volunteers to share ideas about the Key Question. The goal is primarily eliciting and probing student ideas here. When students share their answers, do not indicate if their answers are right or wrong. Make it clear that students are expressing ideas, but those ideas may change based on what they will see in the lessons. Challenge any ideas that are inconsistent with learning from previous lessons.

Activity 1: Analyzing Scientists' Explanation About Animal Growth and Repair

Scientists tell others about new research findings by presenting them at meetings and publishing them in scientific journals. In order to get their work published in well-respected journals, scientists must state their conclusions and explain them with evidence and logical arguments. However, scientists often don't use the terms *evidence* or *science ideas* in their papers because other scientists know how to recognize the elements of an explanation. This activity will help you learn to recognize these elements in a published paper (and sometimes to recognize when elements are missing).

Procedures and Questions

1. Read the simplified paper below and answer the questions that follow with your group.

Taking an Amino Acid Supplement Causes Elderly Men to Increase Protein Production

Introduction

During aging, a gradual decrease in skeletal muscle mass occurs in both rodents (like mice and rats) and humans. This decrease in muscle mass often makes elderly people weaker and affects their ability to move around and their health. Prior studies have shown that when elderly rats were fed high amounts of an amino acid supplement, the rats had more muscle mass and produced more muscle protein (myosin) than similar rats that did not get the amino acid supplement. Scientists wanted to know if the amino acid supplement would also help elderly men produce more muscle protein and if that could stop or slow down the decrease in muscle mass.

Methods

A group of 70-year-old male volunteers were randomly assigned to either an **experimental** or a **control** group. The volunteer men were approximately the same size and weight and in good health. "The nature and potential risks of the study were fully explained to each volunteer and written informed consent was obtained before the study from each participant" (p. 306).

The methods were approved by the local ethical committee. The experimental group was fed complete meals (containing proteins, carbohydrates, and fats) plus an amino acid supplement. The control group was fed the complete meals without the amino acid supplement.

The amount of myosin protein produced was measured by giving all volunteers a tiny amount of labeled phenylalanine (an amino acid found in myosin) and measuring how much myosin-containing labeled phenylalanine each man's body produced. High amounts of labeled myosin would indicate high levels of protein synthesis, whereas low amounts of labeled myosin would indicate low levels of protein synthesis.

Teacher Talk and Actions	
Activity 1	**(20 min)**
Have students read the paper and work independently to answer the questions, which ask students to reflect on how engaging in explanation practices contributes not only to their own understanding but also to their accumulation of scientific knowledge over time.	

Results

Table 1. Myosin Production in the Presence and in the Absence of Amino Acid Supplement

	What They Ate	Amount of Labeled Myosin (Muscle Protein)
Experimental Group	Complete meals plus amino acid supplement	High
Control Group	Complete meals	Very low

Discussion and Conclusion

Our study was designed to assess the impact of amino-acid-supplemented meals on protein synthesis in elderly volunteers. We compared the amount of labeled myosin produced by volunteers fed complete meals plus an amino acid supplement with that produced by volunteers fed only the complete meals. We showed that supplementing a complete diet with the amino acid increased muscle protein synthesis. Additional experiments would be necessary to find out whether increasing protein synthesis will actually reduce or prevent the decline in muscle mass of elderly people.

Source: Rieu, I., M. Balange, C. Sornet, C. Giraudet, E. Pujos, J. Grizard, L. Mosoni, and D. Dardevet. 2006. Leucine supplementation improves muscle protein synthesis in elderly men independently of hyperaminoacidaemia. *Journal of Physiology* 575 (1): 305–315.

a. What is the authors' conclusion?

Taking an amino acid supplement causes elderly men to increase muscle protein production.

b. What evidence do the authors present?

Volunteers who ate complete meals plus an amino acid supplement produced more muscle protein than volunteers who ate complete meals without the supplement.

c. Why do you think the authors included the fact that the volunteers were approximately the same size and weight and in good health in the paper?

The authors were providing evidence that the groups were the same in other ways that might affect their ability to make proteins. For example, if the control group wasn't as healthy as the treatment group that could explain why the control group made very little protein compared with the treatment group. By including the statement the authors were providing evidence that the only difference between the treatment and control group is that the treatment group ate the amino acid supplement.

Teacher Talk and Actions
Activity 1 (cont)
Circulate and help students as needed. Question 1a: Students should realize that the title of the paper is the conclusion. Question 1b: Students should be able to summarize the information in Table 1. Question 1c: If students have difficulty coming up with a reason, ask what someone might wonder if the statement weren't included and point out that anticipating and refuting counter claims is an important part of a scientific explanation.

d. The simplified paper left out a lot of information about the methods that the scientists used in their study. The original paper included details about what participants were fed, where the supplements were purchased, and how the amount of labeled myosin was determined. What do you think is the purpose of including all that information in the original paper?

Evidence used to support a conclusion must be verifiable (able to be checked by others). Providing all the details of the methods would enable other scientists to repeat the experiment to check to see if the results could be verified.

e. The simplified paper also left out references to earlier studies. For example, the original paper cited previous findings showing that when elderly rats were fed high amounts of an amino acid supplement, the rats had more muscle mass and produced more muscle protein (myosin) than similar rats that did not get the same supplement. What do you think is the purpose of mentioning the studies with rats?

The simplified paper only provides evidence that the amino acid supplement caused an increase in the amount of protein the men made. It does not provide evidence that the supplement affected the amount of muscle mass the men had. The studies with rats provided evidence that the supplement also affected the amount of muscle mass the rats had. Because humans are similar to rats in some ways, it is possible that the supplement could have a similar effect on human muscle mass.

Additional point for more advanced students: This is an example of using models to give us ideas about how something might work. The rat is being used as a "model" of a human. Experiments that might not be possible to do with people could be done with rats.

Teacher Talk and Actions
Activity 1 (cont)
Question 1d: This question may be too sophisticated for some students. Those who have participated in science fairs will probably know that scientific experiments need to be replicable. Question 1e: This question links the science ideas that students have been citing in their explanations to the facts and ideas that scientists cite in their papers.

Pulling It Together

1. After examining the scientists' explanation, how would you answer the Key Question, **How are the explanations we have been writing similar to the explanations that scientists write when they publish their work?** Be sure to talk about the criteria that both you and scientists would use for evaluating the quality of an explanation.

When we write explanations we have to support them with evidence, just like scientists do when they write explanations. Our evidence can come from observations we make and from inferences we make from data that can be verified by others, just like the evidence that scientists use. Our explanations have to be consistent with all the evidence.

We also have to use science ideas. The simplified paper we looked at in Activity 1 didn't cite science ideas, but the complete paper did cite references to facts and ideas that had been established in other scientific papers.

Teacher Talk and Actions
Pulling It Together **(5 min)**
Invite students to share ideas about Pulling It Together Question 1, which returns to the lesson Key Question. Direct students to look back at their initial ideas about the Key Question and then consider how their ideas have changed.

2. A friend says in class that chemistry has nothing to do with biology. He says that "chemical reactions only happen in labs, not in people or animals!" After some discussion, the class comes up with the statements below. In the space next to each statement, provide evidence, science ideas, or models to support it.

Statements that will help explain your answer	Evidence, science idea, or modeling activity that supports each statement
Animals increase in mass as they grow.	Data in Table 4.1 provides evidence that a dog, human, lobster, and lizard increase in mass as they grow.
Animal body structures are made up mostly of protein.	Data in Table 4.2 provides evidence that the body structures of fish, shrimp, cow, chicken, and so on are made up of more protein than carbohydrate or fat.
Animals make body structures from their food.	Experiments with labeled carbon atoms provide evidence that herring fish can incorporate atoms from brine shrimp proteins into their bodies, and that 70-year-old men build protein from amino acids in their food.
The proteins that make up animal body structures are different from the proteins in their food.	Data in Table 4.4 provides evidence that keratin (protein making up snake's skin) is a different substance from ovalbumin (protein making up a snake's food).
Chemical reactions are required to make new protein molecules for building body structures.	To make body proteins from food, chemical reactions occur during which atoms rearrange (Science Idea #12).
It is possible to make a body protein by digesting a food protein to amino acids and reordering the amino acids to make a different protein.	The modeling activity where we digested a piece of brine shrimp protein to 5 amino acids and then reacted the amino acids to form a different protein and water molecules showed that it is possible to make a body protein by rearranging amino acids from a food's proteins.

Teacher Talk and Actions

Pulling It Together (cont)

Pulling It Together Question 2 asks students to link evidence and science ideas to statements.

Challenge students to list as much evidence as they can, particularly for the statement that animals make body structures from their food.

Challenge students to identify examples from their own experience that could serve as evidence that animals make proteins that are different from proteins they eat. In addition to the examples listed in the table, here are some others: Ticks and mosquitoes eat blood (mostly hemoglobin) but make body structures made out of other proteins. Table 4.2 shows that an insect (locust) is mostly protein, so students could infer that so is a mosquito and a tick. Because mosquitos and ticks don't dissolve in water, their body structures are probably not made up of hemoglobin.

Closure and Link (2 min)

In Chapter 4, we applied our knowledge of chemical reactions to animal growth and repair. In the next and final lesson of the unit, you will use all that you have learned so far to answer a question about a living thing that is not a plant or an animal.

Lesson Guide
Chapter 4, Lesson 4.5
Explaining Growth in All Living Things

Focus: This lesson gives students a chance to synthesize what they have learned about plant and animal growth and to apply their understanding to a new type of living thing: mushrooms. It is intended to help students generalize the ideas they have learned to all life.

Key Question: How does what we have learned about the growth of plants and animals help us think about the growth of all living things?

Target Idea(s) Addressed

- Students should be able to use all the science ideas from Chapters 1, 2, 3, and 4.

Science Practice(s) Addressed

- Apply scientific ideas, principles, or evidence to construct an explanation for real-world phenomena.
- Read scientific texts adapted for classroom use to determine the central ideas or obtain scientific or technical information to describe patterns in or evidence about the natural world.

Materials

Activity 1: Venn diagram worksheet (from the *THSB* website), Tape, Projector, Slide show: *Plant Growth and Animal Growth*

Activity 2: Projector and Slide show: *Mushrooms*

Advance Preparation

Review student work in Chapters 1, 2, 3, and 4 and be prepared to help students find relevant evidence and ideas in their notebooks.

Phenomena, Data, or Models	Intended Observations	Purpose	Rationale or Notes
Activity 1 Slide comparing German shepherd before and after several months and slide comparing a seedling before and after many years	As it grew, the puppy made more skin and fur, more muscle, and probably more bone. As it grew, the seedling made new leaves, bark, and the tissue that makes up its branches.	To remind students about observations of animal and plant growth at the macroscopic level that they made in earlier lessons.	Though plants and animals don't look much alike, students have learned that they have many similarities in what is going on at the molecular level.
Activity 2 Photos and information about mushrooms	Mushrooms are made up largely of a carbohydrate polymer. Mushrooms get the monomers they need by absorbing them from decaying biological materials.	Students use photos as a new context to which they can apply their knowledge about plant and animal growth to generalize ideas about growth of living things.	Mushrooms were selected because they are familiar, multicellular organisms that are neither plants nor animals. Basic ideas about chemical reactions and conservation in growth apply the same way to mushrooms as to plants and animals.

Lesson 4.5—Explaining Growth in All Living Things

What do we know and what are we trying to find out?

In Chapters 3 and 4, you investigated how plants and animals grow and applied ideas about atoms and molecules to explain how they produce new substances and increase in mass. This final lesson gives you a chance to identify things that plant growth and animal growth have in common and to apply your understanding to a different type of living thing: mushrooms.

> **Key Question: How does what we have learned about the growth of plants and animals help us think about the growth of all living things?**

Teacher Talk and Actions
What do we know and what are we trying to find out? (1 min)
Read the introductory text aloud to your class, including the Key Question. Do not elicit student ideas for this Key Question—they will answer it in the following activities.

Activity 1: Venn Diagram: Animal Growth and Plant Growth

Some things about animal and plant growth are similar, while some things are different. In this activity, you will create a visual, called a Venn diagram, which will help you summarize your ideas about these similarities and differences.

Procedures and Questions

1. Review the before and after photos of animal growth and plant growth that your teacher shows you.
2. Use the template provided by your teacher to create a Venn diagram. Work on your own.
    a. Include as many similarities and differences as you can. You should have at least five items for both, and at least three items each for "only plant growth" and "only animal growth."
    b. Think about both things you can see and things you can't see with the human eye.
    c. Include what happens to mass and what happens to atoms in each of the processes.
    d. Feel free to look through your whole notebook for ideas.

Teacher Talk and Actions	
Activity 1	**(15 min)**

Project the before/after photos of animal growth and plant growth—leave this up as students work.

Hand out the Venn diagram handouts to each student.

Read the text under Step 2 aloud to your class.

Have students work independently to generate their Venn diagrams.

Circulate and help students review their notes. If students are writing statements that are too general, ask them to be more specific, so they can list more similarities and differences.

Science Note to Teacher:
Plants use glucose and N atoms from soil minerals to make amino acids. This enables them to make the proteins they need.

Tape Venn diagram here.

Both:

- *Involve chemical reactions in which substances form that have different properties from the starting substances*
- *Involve the rearrangement of atoms from molecules of starting substances (reactants) to form molecules of ending substances (products)*
- *Involve forming polymers and water molecules from monomers*
- *Total mass is conserved because atoms are not created or destroyed*
- *Measured mass of the growing tree increases just as measured mass of the growing animal/dog increases*
- *The increase in measured mass of both is explained by atoms from outside the system entering the system and reacting with other atoms, which causes the atoms to be incorporated into the plant or animal (system). In both cases, there are now more atoms in the system being measured than there were before. However, mass is still conserved because the mass (number of atoms) outside the plant or animal has decreased by the same amount.*

Animal growth:

- *Growth involves making mainly protein polymers.*
- *Several different amino acid monomers are used to make protein polymers.*
- *Amino acid monomers are made up (mainly) of C, H, O, and N atoms.*
- *The monomers come from food taken in by animals.*

Plant growth:

- *Growth iInvolves making mainly carbohydrate polymers.*
- *One type of monomer (glucose) is used to make carbohydrate polymers.*
- *Glucose monomers are made up of C, H, and O atoms (not N atoms).*
- *Plants make their own glucose monomers, rather than taking them in from their environment.*

SE p. 178

Activity 2: Growth in Mushrooms

In this activity, you will apply everything that you have learned about chemical reactions and growth in animals and plants to a different kind of living organism: mushrooms.

Procedures and Questions

1. Read about mushrooms, observe the photos your teacher will show you, and work with your group to answer the question in Step 1a. Then, answer the question in Step 1b individually.

- Some living organisms are neither plants nor animals. You have probably heard of other groups of living things, such as bacteria, algae, and fungi. Fungi are a diverse group of living things, including a wide variety of mushrooms.
- Mushrooms do not carry out photosynthesis like plants do. They typically get the monomers they need to build their body structures from dead or decaying organisms or other biological materials. They can digest the dead organisms and absorb the monomers. That is why mushrooms are often found attached to the trunk of a dead or dying tree. Trees are made up mostly of cellulose.
- Most mushrooms have the same basic body structures, including a cap that produces spores for reproduction, a stem that holds up the cap, and a threadlike feeding network that takes in nutrients. Mushroom body structures are made up largely of chitin (pronounced "ki-tin") polymers. Chitin is another type of carbohydrate polymer like the cellulose and starch you saw earlier.

   a. Make a list of chemical reactions that could help explain how a mushroom grows and gains mass. Next to each statement, list evidence and science ideas from Chapters 3 and 4 that support each statement. Use the table on the next page to organize your thinking.

Teacher Talk and Actions	
Activity 2	**(12 min)**

Read the text underneath Activity 2 aloud to your class.

Project the slide show *Mushrooms* and leave it up while you read the text under Step 1 aloud to the class.

Small groups: Give students time to work through Step 1a.

Science Note to Teacher:
Chitin is a carbohydrate polymer, but it contains some nitrogen atoms—the structure is not shown to students because it might be confusing as the unit does not cover how plants can use a source of nitrogen to make amino acids or chitin from glucose.

Mushrooms often release digestive enzymes into the wood to help break down large molecules like cellulose to monomers so they can be absorbed.

Fungi function as *decomposers*—if your students are very familiar and comfortable with this term, use it—if not, avoid using it.

Statements about chemical reactions that could help explain how a mushroom grows and gains mass	Evidence and science ideas that support each statement
Like animals, the mushroom digests the cellulose that makes up the tree to glucose/monomers.	Animals digest protein polymers to amino acid monomers (Science Idea #16). OR Evidence: The information is given in the text.
The mushroom probably uses glucose monomers to build chitin.	Animals can build protein polymers from amino acid monomers (Science Idea #15). OR Plants can build carbohydrate polymers from glucose monomers (Science Idea #11).
The mushroom gains mass because the glucose monomers came from outside its body.	The increase in an animal's mass is due to the incorporation of atoms from outside the animal's body (Science Idea #17).
Chitin is a carbohydrate polymer that makes up mushroom body structures.	Evidence: The information is given in the text.

Teacher Talk and Actions
Activity 2 (cont)
Encourage group members to help each other and feel free to help guide students' conversations and thinking.

b. If mushrooms are growing on a fallen dead tree, what do you predict will happen to the mass of the fallen tree as the mushrooms continue to grow? Will the mass of the fallen tree increase, decrease, or stay the same? Does your prediction violate ideas about conservation of matter? Use science ideas from anywhere in the unit to explain your prediction.

The mass of the fallen tree will decrease, because:

- *The fallen tree is an open system, so glucose/monomers that were once part of the cellulose making up the tree leave the tree and enter the mushroom.*

My prediction does not violate ideas about conservation, because:

- *Atoms are not created or destroyed in chemical reactions (Science Idea #7), and*
- *the mass of an atom does not change during a chemical reaction (Science Idea #8), so*
- *the total mass doesn't change in a chemical reaction (Science Idea #9).*
- *If measured mass decreases during chemical reactions, it is because atoms have left the system (Science Idea #10).*
- *The mushroom digested the tree to glucose during a chemical reaction (given in the question).*
- *During this chemical reaction atoms of the tree were not destroyed (use of Science Idea #7) and their mass didn't change (use of Science Idea #8), so the total mass all the atoms involved didn't change (use of Science Idea #9).*
- *The decrease in the measured mass of the tree is because atoms left the tree and became part of the mushroom (use of Science Idea #10).*

Teacher Talk and Actions
Activity 2 (cont)
Step 1b requires students to generalize from earlier evidence from plants and animals. No evidence (specific to mushrooms) is presented, but students can use science ideas to reason about what likely occurs.

Pulling It Together

1. Now that you have completed the unit, how would you answer the Key Question, **How does what we have learned about the growth of plants and animals help us think about the growth of all living things?**

Plants and animals both carry out chemical reactions to produce polymers (e.g., proteins, carbohydrates) needed to build their body structures. Perhaps all organisms do so as well.

During these chemical reactions, atoms of reactants are rearranged to form products. Atoms are not created or destroyed during chemical reactions, so if measured mass of a plant or animal increases, the atoms that become incorporated into its body must have come from outside the organism. For plants, the atoms come mainly from CO_2 in the air. For animals, the atoms come from molecules making up their food.

2. How are changes in the matter that makes up living and nonliving things similar?

Changes in both living and nonliving things can involve chemical reactions. When chemical reactions occur, in either living or nonliving systems, new substances are formed from the starting substances because atoms of molecules of reactants rearrange to form molecules of products.

Atoms (and therefore matter) are not created or destroyed when chemical reactions occur in both living and nonliving things.

Measured mass can increase when chemical reactions occur in open systems, regardless of whether the system is an unsealed plastic bottle or the body of a living thing because atoms of reactant molecules can enter the system and react to form products. This is how living things grow.

Teacher Talk and Actions

Pulling It Together and Unit Closure **(7 min)**

Individuals:

Give students time to review what they wrote in Chapter 1, Lesson 1.1 and answer Pulling It Together Question 1. Encourage students to be reflective about what evidence they've seen that might have helped change their ideas. Students should be able to generalize the following ideas from plants and animals to all living things:

- All living things carry out chemical reactions to produce polymers (e.g., proteins, carbohydrates) needed to build their body structures.
- During these chemical reactions, atoms of reactants are rearranged to form products.
- Atoms are not created or destroyed during chemical reactions, so if measured mass of a growing organism increases, the atoms that become incorporated into the organism's body must have come from outside the organism (e.g., food, air).

Whole class:

This unit showed us how a little bit of basic chemistry can go a long way in answering biological questions. When you get to high school biology, you will see many more ways that chemistry can help us understand what goes on inside living things.